Radar and Satellite Weather Interpretation for Pilots

Radar and Satellite Weather Interpretation for Pilots

Terry T. Lankford

McGraw-Hill

New York Chicago San Francisco Lisbon London Madrid
Mexico City Milan New Delhi San Juan Seoul
Singapore Sydney Toronto

Cataloging-in-Publication Data is on file with the Library of Congress

McGraw-Hill

A Division of The McGraw·Hill Companies

Copyright © 2002 by The McGraw-Hill Companies, Inc. All rights reserved. Printed in the United States of America. Except as permitted under the United States Copyright Act of 1976, no part of this publication may be reproduced or distributed in any form or by any means, or stored in a data base or retrieval system, without the prior written permission of the publisher.

1 2 3 4 5 6 7 8 9 0   DOC/DOC   0 7 6 5 4 3 2

ISBN 0-07-139118-5

The sponsoring editor for this book was Shelley Ingram Carr, the editing supervisor was Daina Penikas, and the production supervisor was Pamela A. Pelton. It was set in Slimbach per the PFS2 design by Kim Sheran and Wayne Palmer of McGraw-Hill Professional's Hightstown, N.J. composition unit.

Printed and bound by R. R. Donnelley & Sons Company.

 This book is printed on recycled, acid-free paper containing a minimum of 50 percent recycled, de-inked fiber.

McGraw-Hill books are available at special quantity discounts to use as premiums and sales promotions, or for use in corporate training programs. For more information, please write to the Director of Special Sales, Professional Publishing, McGraw-Hill, Two Penn Plaza, New York, NY 10121-2298. Or contact your local bookstore.

Contents

Introduction

Throughout the decade of the 1990s, report after report, from both government and industry, has recommended improved aviation weather education for pilots, dispatchers, controllers, and forecasters—both users and providers of weather information. The Federal Aviation Administration (FAA) acknowledges that training is critical to enabling the aviation community to make the best use of weather information to make sound operational decisions and ensure the safety and efficiency of flight. However, with a few notable exceptions at the Aircraft Owners and Pilots Association's (AOPA's) Air Safety Foundation and the National Aeronautics and Space Administration (NASA), little has been accomplished.

New aviation products and services, once the sole domain of the National Weather Service (NWS), FAA, and the scientific community are now directly available, in real time, to the pilot. Among these are weather radar products and weather satellite images. Both weather radar and weather satellite interpretation are sciences within themselves. However, a basic understanding of the capabilities and limitations of these products can assist pilots in making intelligent flight decisions and avoiding weather hazards.

I've been an active general aviation pilot since 1966. Prior to my retirement in 1998 from the FAA as a Flight Service Station (FSS) Controller, I've had the opportunity to attend a half-dozen classes on weather satellite and weather radar interpretation. Some of these classes qualified me to teach these subjects. Additionally, I have over 20 years' experience working directly with radar and satellite products. This has provided me with a unique perspective on how pilots can interpret and apply these valuable resources in an operational aviation environment.

When it comes to weather, Sir William Napier Shaw's *Manual of Meteorology* published in late 1920s nicely sums it up: "Every theory of the course of events in nature is necessarily based on some process of simplification of the phenomena and is to some extent therefore a fairy tale." Often in a discussion of the weather, especially for the nonmeteorologist, we need to simplify the explanation. Additionally, various experts may have different opinions about weather phenomena. Consider, for example, the temperature most favorable for structural icing. You'll see in various texts ranges from +2°C to -20°C. However, the most significant icing range is -2°C to -15°C. With the dynamic nature of weather, at times significant icing can occur outside of either of these ranges. Another example is the most likely temperature to expect a lightning stroke. Some studies show the most active area to be between +11°C and -6°C; other rules of thumb indicate temperatures within 8°C of the freezing level. This, however, does not mean that a lightning stroke cannot occur at warmer or colder temperatures, just that these are the most favorable temperature ranges. These are not inconsistencies, but rather differences of opinion and generalizations, not to be confused with hard-and-fast rules. In fact, in meteorology there are very few absolutes.

Many have touted the necessity for pilots to get the "big picture." This refers to obtaining a comprehensive weather synopsis—that is, the position and movement of weather-producing systems and those that pose a hazard to flight operations. This is important, but it is only one element needed for an informed weather decision. I prefer the term *complete picture*. As well as a thorough knowledge of the synopsis, pilots must obtain and understand all the information available from current weather to forecast conditions. Each report, chart, or product provides a clue to the "complete picture." Each must be translated and interpreted with a knowledge of its scope and limitations. Then, with a knowledge of the complete picture, we can apply the information to a specific flight. As observed by the United States Supreme Court: "Safe does not mean risk free." With the "complete picture," a knowledge of our aircraft and its equipment, and ourselves and our passengers we're ready to assess and manage risk.

Accident reports and commentaries frequently refer to a pilot's poor judgment, namely the failure to reach a sound decision. Pilot judgment is based on training and experience. Training is knowledge imparted during certification, flight reviews, seminars, and literature reading; experience can be best defined as: "When the test comes before the lesson." Unfortunately, failure can be fatal. Pilot applicants have only their instructor to prepare them to make competent flight decisions.

Throughout the book we will use statistics and case studies. With respect to accident statistics and scenarios, some may say: "Why emphasize the negative?" Our goal, along with the Federal Aviation Administration's is accident prevention. People learn through either training or experience. We hope, through a review of various incidents and accidents, to help prevent pilots from becoming statistics. To this end we've used NASA's Aviation Safety Reporting System (ASRS), http://asrs.arc.nasa.gov. ASRS reports are contained in NASA's publication *Callback*. Throughout the book we may refer to *Callback* conversations with the reporter.

While we're on the subject of accidents, let's concede that hindsight is always 20-20. It's easy to sit back in our favorite armchair and analyze and criticize someone else's performance and decisions. When we review an incident or accident in this book, or any other publication or forum for that matter, let's not judge or attempt to blame. Our goal is prevention through education. Therefore, if the reader perceives any judgment or blame in any of the scenarios discussed, it is strictly unintentional; that is not our objective or purpose.

Weather reports and forecasts not associated with radar and satellite products are not within the scope of this book. A complete discussion of these products is contained in McGraw-Hill's companion publication *The Pilot's Guide to Weather Reports, Forecasts, and Flight Planning.*

The first three chapters of this book prepare the pilot to apply the wealth of information available from the National Weather Service's radar network. But, like satellite imagery, radar has limitations. Failure to understand these limitation has, unfortunately, led to a number of tragic accidents. The reader will be taken through the steps necessary to apply radar technology and apply it to the decision-making process.

A basic knowledge of weather radar principles is essential to understanding what radar displays, or more importantly what radar does not display. These subjects are discussed in Chap. 1. The chapter begins with a discussion of how radar sees precipitation and the radar beam. A basic understanding of this subject will put to rest many radar misconceptions and myths. A subject unto itself is radar intensity levels. It may surprise many that the NWS, FAA, and airborne radars do not necessarily display the same intensity levels. This can, and has, led to pilot-controller misunderstanding, and has the potential for disaster. As important as intensity levels are radar limitations. As yet radar does not directly see turbulence. However, there are certain correlations that can be made between radar intensity levels and turbulence.

The final section of Chap. 1 concerns weather avoidance. If the reader comes away with only one lesson, it should be that radar and other storm detection systems are for weather avoidance, not weather penetration! The discussion includes various methods that allow a pilot to avoid hazardous weather.

Pilots have access to various weather radar displays, both during preflight planning and in flight. These may be in the from NWS radar, Air Traffic Control (ATC) radar, or airborne weather radar. These are presented in Chap. 2. Each type of radar and presentation has it own advantages and limitations. These must be understood for safe and efficient flight operations.

NWS radars have the power to penetrate weather, with a wavelength and beamwidth specifically designed to detect precipitation. Even so, they have limitations of ground clutter and can be blocked by terrain. There is always a time lag between the time the observation is made, the data is processed, and information is displayed to the pilot.

ATC radars are intended to separate aircraft, not display weather. This is not to say that ATC radars do not have limited weather capability. But during poor weather the controller may select various circuits to eliminate or reduce weather returns. Each of these will be discussed along with the weather that these radars can display.

Lightning detection equipment, trade named Stormscope, was invented in the mid 1970s by Paul A. Ryan as a low-cost alternative to radar. Stormscope and similar lightning detectors sense and display electrical discharges in the approximate range and azimuth of the aircraft. Lightning detection systems are discussed in the final section of Chap. 2. Like radar and satellite, lightning detection systems have advantages and limitations.

Pilots have access to weather radar data in the form of a conventional radar scope, composite radar images, graphic products, and textual messages. Chapter 3 discusses these weather radar products. Each system or product has a specific purpose and its own applications and limitations. Of these radar products, pilots are most familiar with the National Radar Summary Chart and Convective SIGMETs. In addition to these products, pilots have access to Radar Weather Reports (RAREPs), also called SDs (for storm detection); the National Reflectivity Mosaic, Radar Coded Message Images, Alert Weather Watches (AWWs) and Weather Watch Bulletins (WWs). Additionally, some Center Weather Advisories (CWA) relay information on hazardous convective activity.

Since 1960, weather satellites have been orbiting the earth. Chapters 4, 5, and 6 are devoted to current, state-of-the-art, satellite imagery available to pilots. Like every other weather report or forecast, satellite imagery has limitations. These limitations must be understood for proper application. The reader will be taken through, in a logical order, the steps necessary to apply this valuable, but often overlooked, tool.

Chapter 4 begins with an introduction to satellite imagery. Without this basic understanding it is difficult, if not impossible, to effectively apply this technology to real-time flight operations.

There are two basic types of meteorological satellites: polar orbiters and equator orbiters. Pilots usually have access to the equator orbiters— the Geostationary Operational Environmental Satellites (GOES). GOES orbit the Earth once every 24 hours. Therefore, from the satellites' view the earth appears to remain stationary.

There are three basic types of satellite images: visible, infrared (IR), and water vapor. A visible image is a snapshot from space; infrared is a temperature measurement of the Earth's surface or cloud tops; water vapor provides a display of moisture patterns. For our purposes we will discuss only visible and IR images. Infrared images may be enhanced to provide better detail of specific phenomena. To properly apply an enhanced IR image, the user (pilot) must understand how the image was altered. Much information can be gained when comparing two images for the same time period, or by viewing a series of successive images known as a *loop*.

Chapter 5 continues the discussion with geographical and weather features seen on satellite imagery. Geographical features include terrain features and cultural boundaries—local, state, and international borders. As well as cultural boundaries, major lakes are often part of the "grid" that helps to identify geographical areas. Terrain features can be distinct or subtle, and both visible and IR images are often needed to analyze them. Additionally, the time of day has an effect on the satellite's depiction of Earth's features.

Next we move on to weather features seen on satellite imagery. Pilots will learn to determine if clouds are high, middle, or low, thick or thin. This, again, often requires a comparison of both visible and IR images. Texture (lumpy, flat, or fibrous) allows us to determine cloud type. The subtle difference between snow and cloud cover is discussed. A significant application of satellite imagery is a determination of areal coverage. We can determine if an area of clouds and weather is circumnavigable, and if so, which is the best route. The satellite can reveal the location and movement of major weather systems, such as high- and low-pressure areas, fronts, squall lines, and the jet stream, as well as the extent of local features, such as dust, smoke, fog, and volcanic ash.

Chapter 6 applies the knowledge gained from the previous two chapters into practical application. The discussion relates satellite features to aviation weather hazards. This in turn provides the pilot with valuable decision-making information. A pilot might decide to penetrate, circumnavigate, or cancel a flight on the basis of training, experience, the aircraft and its equipment, and personal minimums.

The final chapters combine satellite and radar interpretation. We will show, in practical application, how these products apply to flight planning and risk assessment.

In Chap. 7, we review the steps necessary to assemble the information necessary for a safe and efficient flight. We begin with flight planning. Flight planning consists of evaluating all elements physically related to the flight, such as terrain, altitudes, and the environment. With an evaluation of terrain and altitudes complete, we can move on to an assessment of the environment—weather, personal minimums, and alternates. The first part of the environmental evaluation is obtaining the complete picture through a preflight weather briefing. This knowledge can then be applied to our personal minimums and available alternates. An assessment of terrain, altitudes, and environment is the first step in the go–no go decision.

With the knowledge gained from previous chapters, we can now apply them to real-world concerns. Here the pilot will learn how to apply the various satellite and radar products, and avoid the pitfalls of their limitations. Chapter 8 delves into the often ambiguous subject of risk assessment and management. Risk assessment and management begins with flight planning strategies, including personal minimums, preparation, and evaluation. With this accomplished we can move on to risk evaluation. In the final sections we apply flight planning strategies and risk assessment, using actual weather scenarios, to apply risk management techniques.

Appendix A, "Weather Report and Forecast Contractions," provides the reader with a quick-reference guide to the contractions used on various weather products. This appendix was included to assist the reader in decoding various contractions, which are not directly within the scope of this text.

Appendix B, "NWS Weather Radar Chart and Locations," contains a graphic display of the location of NWS radars and a decode list of their location identifiers.

Appendix C, "Weather Advisory Plotting Chart and Locations," provides a reference for the location description used on Convective SIGMETs and radar-derived Center Weather Advisories.

Throughout the book we mention various geographical areas. To assist the reader these locations are graphically depicted in Appendix D, "Geographical Area Designators." These are the same designators used in the Area Forecast (FA) and many other aviation weather products.

A Glossary is included. The Glossary serves as a quick-reference guide for terms and concepts.

The following chapters explain, hopefully with a little humor, how to use radar and satellite products, then apply them to a flight situation. Throughout the text we've poked some fun at ourselves and others, especially government agencies like the FAA and NWS, since sometimes we tend to take ourselves a little too seriously. By doing this we've tried to inject a "reality check." However, incidents are not intended to disparage or malign any individual, group, or organization; the sole purpose is illustration.

Great effort has been made to insure information is current and accurate, as of the time of writing. Unfortunately, especially in aviation and aviation weather, the only thing that doesn't change is change itself.

Radar and Satellite Weather Interpretation for Pilots

Introduction to Radar

" **A** disastrous thunderstorm accident close to Bowling Green, Kentucky, in 1943 that involved an American (Airlines) DC-3 started a chain of events that eventually led to the first systematic research into thunderstorm behavior. The plane crashed onto the ground either near or under a severe thunderstorm. Buell [C. E. Buell, chief meteorologist, American Airlines 1931–1946] initiated a letter to the Civil Aeronautics Board pointing out the appalling dearth of understanding of what actually occurs inside a thunderstorm, as evidenced by the accident investigation. He recommended a massive research effort be organized to probe into thunderstorms and document their internal structure," according to Peter E. Kraght in *Airline Weather Services 1931–1981*.

Thunderstorms contain just about every weather hazard known to aviation: turbulence, icing, precipitation (including hail), lightning, tornadoes, gusty surface winds (including low-level wind shear), effects on the altimeter, and low ceilings and visibilities. Under certain conditions they can produce high-density altitude. Thunderstorm hazards are illustrated in Fig. 1-1.

It should be no surprise that thunderstorms have the potential to produce severe to extreme turbulence. Vertical motion is the structural

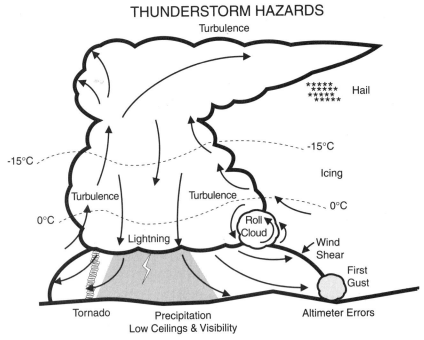

Fig. 1-1. Thunderstorms can produce every kind of aviation weather hazard.

CASE STUDY

Just when you think you've seen all the hazards associated with thunderstorms something different comes up. Take, for example, U.S. Air Force Lt. Col. William Rankin. He was forced to bail out of his jet at 47,000 ft over Virginia. You guessed it, into a raging thunderstorm. As he fell through the storm he plummeted to 10,000 ft before opening his chute. For the next 40 minutes the Colonel rode the storm's up- and downdrafts, with lightning bolts flashing all around. He finally wound up in some trees about 65 miles from where he bailed. His adventure gives a whole new meaning to "sport parachute jumping."

basis of the cell. In thunderstorms, the width of up- and downdrafts may vary from a few feet to several thousand feet. These drafts affect the aircraft's altitude as it flies through the thunderstorm. It is virtually impossible to hold altitude. Altitude changes of several thousand feet are not unusual.

Downdrafts continue below the base of the cloud with significant speed to within 300 to 400 ft of the ground. These drafts constitute a significant hazard to flight beneath the thunderstorm, which is often in heavy rain and poor visibility. The most severe turbulence is encountered most frequently near the freezing level, but can also occur from the ground to above the cloud tops. Significant turbulence can also occur in clear air well away from the cell itself.

Icing is another principle hazard in thunderstorms, expect icing in all storms at elevations above the freezing level. Although thunderstorm clouds are usually limited in diameter, even a short-duration exposure can result in severe icing. The most severe icing can be expected between the freezing level and −15°C, as illustrated in Fig. 1-1.

Hail can be one of the worst hazards of thunderstorm flying. Great amounts of hail and the largest stones generally are found in the larger and taller storms, but many thunderstorms have no hail associated with them. In general, large hail occurs in severe thunderstorms. Frequently hail is carried aloft and tossed out the top or side of the cloud by updrafts, and hail may be encountered in clear air several miles from the cloud. Hail frequently exists in thunderstorms even though not reported at the ground. Flight beneath the anvil should be avoided because of the hail hazard, as illustrated in Fig. 1-1.

The largest hailstones and greatest frequency occur in a mature thunderstorm cell, usually at altitudes between 10,000 and 30,000 ft. Hail usually falls in streaks or swaths beneath the thunderstorm, covering an area about one-half mile long and 5 miles wide. Hail typically occurs in the rain area within the cloud, under the anvil or other overhanging cloud, and up to 4 miles from the cloud. Hail can also include other forms of frozen precipitation with differing origins, such as snow.

Hail can cause severe damage to objects on the ground as well as aircraft in flight. Blunted leading edges, cracked windscreens, and frayed nerves are common results of a hail encounter. There have been a number of multiple turbine engine power-loss and instability occurrences, forced landings, and accidents attributed to operating aircraft in extreme rain or hail. Investigations have revealed that rain or hail concentrations can be amplified significantly through the turbine engine core at high flight speeds and low engine power. Rain or hail may degrade compressor stability, combustion flameout margin, and fuel control run-down margin. Ingestion of extreme quantities of rain or hail through the engine core may ultimately produce engine problems, including surging, power loss, and flameout. Pilots must be familiar with these phenomena and comply with the manufacture's recommendations. Like most thunderstorm hazards, avoidance may be the only safe alternative.

Lightning experienced in a thunderstorm can cause temporary blindness so that control of the aircraft by reference to instruments may be momentarily lost. Damage to navigational and electronic equipment can also create a hazard. Small punctures may result in the aircraft's skin from direct lightning strikes. Lightning is found throughout and adjacent to the thunderstorm cloud, but is most frequent and severe from the freezing level up to about $-10°C$. (We'll discuss lightning in more detail in Chap. 2, "Thunderstorm Avoidance Systems.")

Tornadoes are produced by severe thunderstorms. They are whirlpools of air, cloud, and debris that range in diameter from 100 ft to a half mile. Pressure is extremely low in the center of the small concentrated vortex. Tornado winds probably reach 200 to 300 knots, although damage patterns indicate that winds of over 400 knots can exist. Tornadoes appear as funnel-shaped clouds extending from the base of thunderstorms and usually move at 25 to 50 knots. Their paths range from a few miles to probably less than 50 miles, although squall lines and frontal systems can produce a series of tornadoes that can cover hundreds of miles. Their exact path is erratic and unpredictable.

Gusty and variable surface winds are associated with thunderstorms. Usually the first gust, or gust front, precedes the arrival of the roll cloud and onset of rain as the thunderstorm approaches. Frequently it stirs up dust and debris as it plows along, announcing the thunderstorm's approach. The strength of the first gust frequently is the strongest observed at the surface during a thunderstorm. It may approach 100 knots in extreme cases. The roll cloud is not always present, but is found most frequently on the leading edge of fast-moving thunderstorms. It represents an extremely turbulent area.

Pressure usually falls rapidly with the approach of a thunderstorm, then rises sharply with the onset of the gust front and arrival of the cold downdraft and heavy rain. Pressure then normally falls back as the storm moves on. This cycle of pressure change may occur within 15 minutes. Height indicated on a pressure altimeter during the storm may be in error by over a hundred feet.

Heavy rain brings lowering ceilings and visibilities. With severe storms, ceilings and visibilities can be at or near zero.

Thunderstorms may produce a phenomenon known as a heat burst. Carolyn Kloth, forecaster at the NWS's Aviation Weather Center (AWC), has proposed that a heat burst may be another thunderstorm hazard for aviation: high-density altitude. Heat bursts increase turbulence and wind shear, develop in radar-echo-free air, and affect the pressure altimeter. They typically occur in the dissipating stage of nighttime thunderstorms.

In the effort to reduce thunderstorm hazards, radar was one of the many projects, and played a significant role, in the early 1950s. Early airborne radars had many technical problems, but by the beginning of the 1960s most airliners were equipped with airborne weather radar. Ground-based radar was also developed.

Access to radar information has increased over the years. Until the mid 1990s, NWS radars in the east and FAA radars west of the Rockies, supplemented by a few NWS sites, were used to compile a National Radar Summary Chart. An NWS network in the west was originally thought unjustified because severe weather is relatively rare in that area. Radar Weather Reports (RAREPs) and convective analysis charts are routinely transmitted on NWS and FAA circuits and are available through many private vendors. Many twin- and, more and more, single-engine aircraft are being equipped with airborne weather radar or lightning detection systems, and there have been several attempts to place ground-based radar displays in the aircraft.

The last of the old WSR-57 weather radars have been decommissioned, and replaced by the next generation (NEXRAD) weather radar (WSR-88D). NEXRAD is a Doppler radar system which, with a few minor exceptions, provides coverage from coast to coast in the contiguous United States, Hawaii, Alaska, and the Caribbean.

Like weather observations and forecasts, each system, product, or service has its own particular application and limitations that must be thoroughly understood for safe and efficient flight.

Radar works on the detection of returned electromagnetic energy, or echoes. The transmitter sends out a pulse of energy at the speed of light. As the energy strikes a target, some of the energy is returned to the antenna. Because the antenna scans through the atmosphere, the bearing of the target can be determined; since the pulse moves at the speed of light, range to the target can be established. This information is displayed to the operator on a plan position indicator (PPI). It's important to remember that the PPI presentation does not depict the vertical extent of the storm; it represents only a small cross-section. This is extremely important for pilots using airborne weather radar systems.

In oversimplification terms, radar displays an image dependent on reflected energy or backscatter. Intensity depends on several factors, among them reflecting particle or droplet size, shape, composition, and quantity. The image displayed also depends on a number of factors. These include radar frequency, power output, antenna type and size, and radome. These limitations affect airborne radars to a much greater degree than ground-based units.

Ground-based NEXRAD radars are capable of displaying both precipitation and clouds. However, Radar Weather Reports (RAREPs

AN OVERSIMPLIFICATION

We usually talk about reflected radar energy. This isn't quite true for precipitation. Although there is a small amount of reflected energy, the radar emission causes an oscillation of the electrical charge in the precipitation particle. This causes the particle to generate energy, while much of the radar-transmitted energy passes through the target. This is why distant precipitation can be detected behind other precipitation targets. However, from an operational viewpoint and for simplicity we will refer to this phenomenon as reflected or "backscatter" energy.

or SDs) and the Radar Summary Chart display precipitation only. Therefore, a precipitation-free area on these products does not translate into a cloud-free sky.

As can be seen from Table 1-1, liquid precipitation is a good radar reflector, while frozen precipitation is not. Rain is the best reflector; although wet hail and wet snow—because of their liquid water coating—are also good radar reflectors. Dry hail and dry snow are poor reflectors, while water vapor, small dry hail (because of its small size), and ice crystals provide essentially no radar return.

The Radar Beam

While it's not necessary to be an electronic engineer, a basic understanding of radar will help in the recognition of its uses and limitations.

Radars transmit their signals in brief bursts called *pulses*. The unit of measurement generally used to define pulse length is the microsecond (μs). Pulses range in length from a fraction of a microsecond to several microseconds. As pulse length increases, the strength of the signal increases. Therefore, a long pulse will detect weaker targets than a short pulse. However, the shorter the pulse, the greater the range accuracy. The emission of pulses is timed so that, within the normal range of the radar, the echo from one pulse is returned to the antenna before the next pulse is transmitted.

The size and shape of the radar beam is determined by the size and shape of the antenna and the wavelength of the transmitted energy. The purpose and physical location of the antenna is governed by its size and design. Ground-based units have virtually no limit on size or shape. On the other hand, airborne units, especially for smaller aircraft, have major design constraints.

Resolution describes the ability of the radar to detect discrete targets. Range resolution is the ability of the radar to distinguish between two or more targets at the same azimuth, but at different ranges. For example, range accuracy with a pulse length of 1 μs permits targets separated by as little as one-tenth of

TABLE I-I. Precipitation Reflectivity

GOOD	POOR	NONE
Rain	Dry hail	Water vapor
Wet hail	Dry snow	Small dry hail
Wet snow		Ice crystals

a mile to be resolved as separate echoes. A pulse length of 5 μs would decrease the accuracy to a separation of about one-half mile.

Beam resolution describes the ability of the radar to distinguish between targets at the same range, but at different azimuths. Farther from the antenna, the focused radar beam becomes wider with distance. This is illustrated in Fig. 1-2. When a target is within any portion of the beam, and gives sufficient backscatter, the radar detector interprets this as filling the whole beam. To obtain beam resolution, two targets must be separated by at least one beamwidth. Because of increasing beamwidth with range, beam resolution decreases with increasing range. Notice in Fig. 1-2 that the more distant target appears larger and solid, even though it is actually two separate, smaller targets. Thus, a narrow beam will provide a more representative echo than a wide beam.

Attenuation is any process that reduces power density within the radar beam. A target or obstruction close to the antenna may absorb and scatter so much of the energy that little passes to a more distant target. Weather radars have built-in circuits that compensate for reduced power returns of distant targets. Therefore, within the normal range of the radar, targets at various ranges are displayed in relation to their true intensity.

Significant attenuation can result from heavy rainfall and steep precipitation gradients. This often results from severe weather. Precipitation attenuation is related to wavelength. It can be a serious problem with sets of 5-centimeter (cm) wavelength or less—especially in heavy rain. Rain attenuation is not significant with high-power 10-cm radars. However, attenuation with relatively low-power 3-cm airborne weather radars can be significant.

CASE STUDY

In the early days of radar meteorology, forecasters were puzzled by the appearance of a solid line of echoes in the distance, which routinely broke into individual cells as the system approached the radar; then grew back into a solid line as they moved off into the distance.

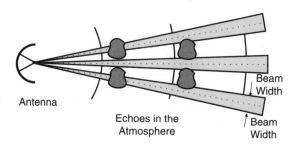

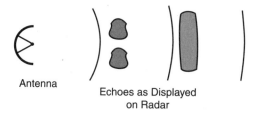

Fig. I-2. Beam resolution describes the ability of the radar to distinguish between targets at the same range, but at different azimuths.

According to the National Transportation Safety Board (NTSB), precipitation attenuation was a contributing factor in the crashes of a Southern Airways DC-9 in 1977 and an Air Wisconsin Metroliner in 1980. Precipitation attenuation is not significant with NWS 10-cm high-power units; however, it can be a serious problem with units of 5 cm or less, especially in heavy rain. The NTSB recommends:"…in the terminal area, comparison of ground returns to weather echoes is a useful technique to identify when attenuation is occurring. Tilt the antenna down and observe ground returns around the radar echo. With very heavy intervening rain, ground returns behind the echo will not be present. This area lacking ground returns is referred to as a shadow and may indicate a larger area of precipitation than is shown on the indicator. Areas of shadowing should be avoided."

In August 1985, a Delta Air Lines L-1011 crashed at Dallas/Fort Worth Airport. The NTSB was unable to determine if the crew had been using airborne weather radar at the time of the crash. The NTSB report did state, however:"The evidence concerning the use of the airborne weather radar at close range was contradictory.

ATTENUATION

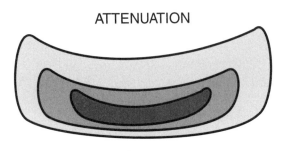

10-cm NWS

3-cm Aircraft

Fig. I-3. Precipitations attenuation, caused by close targets absorbing and scattering the radar's energy, can be a serious problem with low-power, short-wavelength sets.

Figure 1-3 illustrates the effects of precipitation attenuation, showing how a heavy precipitation pattern with a very strong gradient might appear on an NWS 10-cm radar, compared with the same weather system as seen on 3-cm aircraft units. A pilot seeing the pattern on a 3-cm set might elect to penetrate the weather at what appears to be its weakest point only to find the most severe part of the storm, or find additional severe weather where the radar indicated a precipitation-free area.

When you are using an airborne weather radar it is imperative to understand the particular unit, its operational characteristics, and limitations. Just reading through the brochure that comes with the equipment is certainly not enough to prepare a pilot to translate the complex symbology presented on the display into reliable data. A training course with appropriate instructors and simulators is mandatory for safe and efficient operation of the equipment.

Range attenuation is the loss of power density due to distance from the antenna. Power density in the radar beam is attenuated by range and decreases as range increases. To compensate for range attenuation of echoes, a sensitivity time control circuit (STC) is used. When STC is engaged, indicated signal strength displayed on the radar indicator is independent of range; in other words, intensity of the presentation is compensated for change with range. However, because of relatively low power, with larger beamwidth, airborne radar will indicate weaker storms at the maximum range of the radar. Effective range is usually limited to 60 to 80 nm. This is why distant storms may appear to grow in intensity as the aircraft approaches the cell.

Radar Intensity Levels

Prior to the commissioning of the NEXRAD network, weather radar intensity levels were described as Video Integrator and Processor (VIP) levels. This system divided the strength of radar returns into six intensity levels. NEXRAD radars display echo returns as energy described in decibels (dBZ).

FACT
You may still hear pilots and controllers speak of VIP levels, even though VIP no longer exists. It's like pilots who report icing was "heavy." (It seems that the maximum intensity reported for icing was "heavy" untill 1968. That's almost before my time!) Both terms are misnomers.

For purposes of the Radar Weather Reports (RAREPs) or SDs (storm detections) and the Radar Summary Chart, the National Weather Service still assigns six levels. However, instead of being based on the Video Integrator and Processor, they are derived from their dBZ level. Table 1-2 contains Radar Precipitation Intensity Levels.

As mentioned, for purposes of the SD and Radar Summary Chart the NWS uses six levels of intensity. (*Note:* Pilots will see only the intensity symbols on RAREPs or the SD.) At the present, pilots can

TABLE 1-2. Radar Precipitation Intensity Levels

PRECIPITATION INTENSITY	SYMBOL	dBZ	APPROXIMATE WSR-88D COLOR	ASSOCIATED CONVECTIVE WEATHER
1 Light	—	>15	Green	LGT-MOD TURBC PSBL LTNG
2 Moderate	No symbol	30 to 39	Light yellow	LGT-MOD TURBC PSBL LTNG
3 Heavy	+	40 to 44	Dark yellow	SEV TURBC AND LTNG
4 Very heavy	++	45 to 49	Orange–light red	SEV TURBC AND LTNG
5 Intense	X	50 to 54	Red	SEV TURBC LTNG WIND GUST HAIL
6 Extreme	XX	≥55	Dark red–purple	SEV TURBC LTNG EXTENSIVE WIND GUST HAIL

expect Air Traffic Control (ATC) to provide radar intensities using these six levels. These levels do not correspond to displayed airborne weather radar levels. Typically, airborne weather radars display precipitation intensity in three or four levels. These levels generally correspond to NWS levels of 1 through 3 or 4, respectively. More about this in Chap. 2, "Thunderstorm Avoidance Systems."

NWS radar displays, and most commercial vendors, indicate precipitation by using the colors described in Table 1-2. Colors typically range from green (lowest precipitation rate) to red (highest precipitation rate). *A word of caution:* With convective weather, even a green return represents potential danger. This can be seen in the associated convective weather portion of Table 1-2. This is why, as we shall see, that determining the character of a weather system—convective or nonconvective—is extremely important in estimating potential hazards.

Radar Limitations

The most important thing for pilots to understand is that radar sees precipitation, not turbulence or the other hazards associated with thunderstorms. However, there is a direct relationship between precipitation intensity and turbulence. And, although some airborne weather radars do have turbulence detection modes, they can detect only turbulence within areas of precipitation. They cannot detect clear air turbulence, which is often associated with convective activity.

Figure 1-4 contains a turbulence probability chart. The graph shows the percentage of turbulence probability on the vertical scale (0 to 100 percent); the horizontal scale represents radar intensity levels.

Probability is often used in aviation weather. For example, let's take a forecast for a 50 percent probability of turbulence. A pilot flying through an infinite number of such events can expect to encounter turbulence one-half of the time. However, since penetrating an infinite number of events is impossible, a pilot may spend an entire career and never encounter any turbulence; on the other hand, a pilot might encounter a greater number of turbulence events than the probability would tend to indicate. But, on average, the probability holds.

It's important for pilots to understand the concept of probability. Many a pilot has unwarrantedly criticized aviation weather forecasts by failing to understand this notion. The bottom line: When you fly into an area of

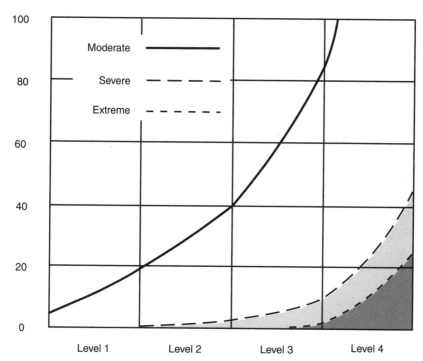

TURBULENCE PROBABILITY

Fig. 1-4. Even though the probability of severe or extreme turbulence in a level 3 echo is small, pilots can never assume penetrating the cell is without hazard.

probability, if you encounter the phenomenon, they're right; if you don't encounter the event, they're still right! Probability allows us to access risk and take appropriate precautionary measures. More about risk evaluation and management in Chap. 8, "Evaluating Risk."

Although it is not directly depicted in Fig. 1-4, pilots can expect an almost 100 percent probability of light turbulence in any area of convective rain. An area of level 1 echoes, not associated with higher intensity levels, has almost no chance of producing severe or extreme turbulence. However, as can be seen from the graph, level 1 intensity echoes produce a 5 to 20 percent probability of moderate turbulence.

Like level 1, level 2 intensity echoes have almost no probability of extreme turbulence. However, the probability of moderate turbulence ranges from 20 to 40 percent. The probability of severe turbulence is low, up to about 5 percent.

DEFINITION

Probability: The ratio of the chances favoring an event to the total number of chances for and against the event.

Figure 1-4 shows a high probability of moderate turbulence for level 3 echoes—40 to 85 percent. The probability of severe turbulence has increased to between 5 and 10 percent, with a slight chance of extreme turbulence, although less than about 2 percent.

Pilots can expect the probability of moderate turbulence with any level 4 echo. The chance of severe turbulence increases to between 10 and 50 percent; extreme turbulence probability increases from greater than about 2 percent to 25 percent.

Another significant airborne hazard is hail. Research indicates damaging hail to be on the order of $3/4$ inch or greater in diameter. Three-quarter-inch hail is the threshold for a severe thunderstorm, along with surface winds of 50 knots or greater. Hail size probability is illustrated in Fig. 1-5. The probability of damaging hail begins with level 3 echoes and increases with echo intensity.

NOTE

With the adoption of the METAR/TAF codes, severe thunderstorms are no longer specifically identified in weather reports and terminal aerodrome forecasts. However, pilots should specifically look for winds of 50 knots or greater and hail $3/4$ inch or greater in these reports—which still define a severe thunderstorm.

Hail smaller than $3/4$ inch in diameter typically does not cause structural damage. Hail (GR) is reported in METAR when the size is equal to, or greater than $1/4$ inch. The specific size of the hail is reported in the remarks of the report. Small hail or snow pellets are reported by using the METAR code GS. The size of small hail or snow pellets are not reported.

CASE STUDY

We were on an IFR flight in a Cessna 172 from Lancaster's Fox Field in California's Mojave Desert to Van Nuys. We flew into a big, black, ugly-looking cloud. In the cloud turbulence was moderate and the sound of hail hitting the windscreen was deafening! I thought at best I would be buying a new windscreen, but it turned out there was no damage at all. According to the turbulence and hail probability charts this was a level 2 cell. I don't do silly things like that anymore.

The bottom line: Although penetrating a level 3 cell may appear to be an acceptable risk, it is not! Ground-based weather radars display level 3 echoes as yellow. Airborne radars show this

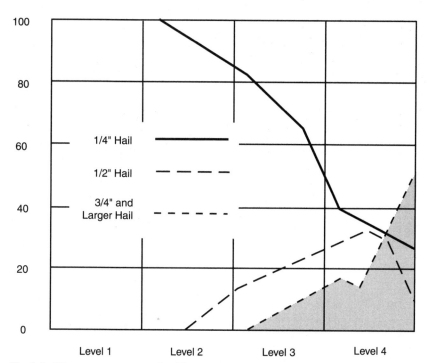

Fig. 1-5. Pilots must treat every level 3 or greater return as extremely dangerous.

intensity level as red. Whenever radar shows a level 3 or greater storm, the entire storm cell should be considered extremely hazardous and must not be penetrated. The only safe way to operate in areas of convective activity is to avoid all level 3 or greater returns. Remember that radar is a storm avoidance, not a storm penetration, device.

Pilots must remember that hazardous areas of turbulence and hail are not necessarily associated with areas of maximum radar echoes. Also, the probabilities of turbulence apply to the whole storm system, not just the maximum radar echo areas. So, what does that mean? Pilots must avoid any intensity level associated with the maximum storm echo intensity.

The violent nature of thunderstorms causes gust fronts, strong updrafts and downdrafts, and wind shear in clear air adjacent to the storm out to 20 miles with severe storms and squall lines. Precipitation, which is detected by radar, generally occurs in the downdraft, while updrafts remain relatively precipitation free. Clear air or the lack of radar echoes does not guarantee a smooth or safe flight in the vicinity of thunderstorms.

Figure 1-6 is an NWS NEXRAD radar image from Flagstaff, Arizona. It illustrates a typical summer afternoon for the southwest United States. These are typical air mass thunderstorms. They are usually circumnavigable visually or with storm avoidance equipment. However, *a word of caution:* They can form in lines or clusters. Note the cells west and north of Prescott (A and B). The outermost portions of the storms are blue. (I know the illustration is black and white, so you'll just have to take my word for it.) Blue indicates cloud or very light rain—approximately −5 to +15 dBZ. The next contours are green and light yellow—level 1 and 2; the inner contours (which appear dark on the black-and-white image) represent levels 3, 4, and higher.

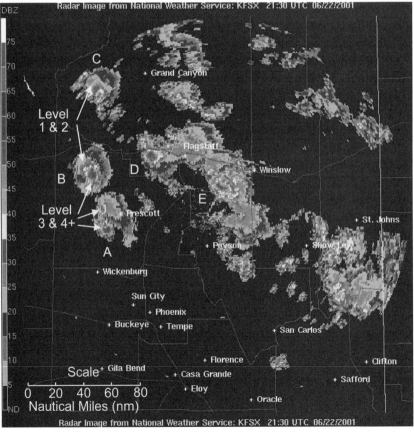

Fig. 1-6. Pilots must avoid level 1 and 2 returns associated with higher-level cores by at least 20 nautical miles.

As we will see, the general rule for storm avoidance is 20 nautical miles (nm). This means a corridor 40 nm wide between storms! If we take the two cells west and just north of Prescott (A and B), the level 1 returns are separated by less than 10 miles. Therefore, a pilot should not attempt to penetrate this area. The distance between the cells northwest of Prescott (B and C) are just 40 nm. A pilot could expect a relatively reasonable ride between these cells.

Earlier steep rain gradients were mentioned. Examples of steep rain gradients are illustrated in Fig. 1-6 at D and E. At D the gradient goes from cloud returns almost directly to level 4 and greater echoes. At E the gradient is not as steep as D, with several miles of level 1 and 2 returns prior to the more intense precipitation—but still significant.

Weather Avoidance

The most severe turbulence is not necessarily found at the same place that gives the greatest radar reflection, although the frequency and severity of turbulence increases with radar reflectivity. Research indicates that the frequency and severity of turbulence encounters decrease slowly with the distance from the storm core. Severe turbulence is often found in tenuous anvil clouds 15 to 20 nm downwind from a severe storm.

Significant and sometimes severe turbulence often extends outside of the storm. The clear air on the inflow side of a storm is a place where severe turbulence often occurs. This is typically the southwest quadrant. At the edge of a cloud, the mixing of cloud and clear air often produces strong temperature gradients associated with rapid variation of vertical velocity, thus significant turbulence. Studies indicate there is no useful correlation between the external visual appearance of thunderstorms and the turbulence and hail within them.

Research indicates a relationship between turbulence above storm tops and the airspeed of upper tropospheric winds. When the winds at storm top exceed 100 knots, there are times when significant turbulence may be expected as much as 10,000 ft above the cloud tops. This value may be decreased 1000 ft for each 10-knots reduction of wind speed. Therefore, pilots should plan to clear thunderstorm tops by 1000 ft for each 10 knots of wind speed at cloud tops.

Turbulence below thunderstorms can also be severe. This is especially true in the dry environment. The dry environment

CASE STUDY

The following Colorado Springs (COS) METAR and UUA is an example of a dry thunderstorm environment.

METAR KCOS 202150Z 36012G24KT 45SM -TSRA BKN090 OVC250 23/08 A3019 RMK TS OVH MOV E OCNL LTGCG SW E

COS UUA/OV COS/TM 2156/...TP PA60/RM AIRSPEED + - 40 KTS, THOUGHT I WAS IN THE TWILIGHT ZONE

A thunderstorm with rain showers, a 15° temperature/dew point spread, strong, gusty surface winds, and lightning are being reported. Add the PIREP and there is a high probability of significant turbulence and wind shear.

CASE STUDY

While cruising at FL330, the aircraft encountered severe turbulence and lost several hundred feet of altitude. There were injuries to two flight attendants and one passenger.

We were deviating around weather on a route suggested by ATC and our own flight dispatcher. We also agreed with their suggestions, as we had a good picture on our airborne radar. Approximately 10 minutes prior to the encounter, we had been visual, but at the actual time of the encounter we were IFR in cirrus-type clouds. At the time of the encounter we were approximately 25 miles from the nearest contouring cell as depicted on our radar. Since we were in clouds and radar showed us to be on a clear path, we can only assume we encountered a wind shear situation or flew into a rapidly developing buildup that did not contain enough moisture to give a return on our radar.

is characterized by typically high cloud bases and low surface relative humidity. Strong outflow winds and severe turbulence often exist below the thunderstorm cloud base.

Severe weather often accompanies a single thunderstorm, or a line or cluster of echoes moving at 40 knots or more. The stronger the radar return, the greater the probability and severity of turbulence and hail—which can extend above, below, and to 20 nm away from the storm. Rapidly growing storms can increase in intensity in minutes and grow at the rate of 7000 ft/min. The most common thunderstorm erratic motions are: right-turning echoes, splitting echoes, and merging echoes. All three are indicators of severe or extreme turbulence and large hail.

Maximum turbulence occurs in areas that contain the most abrupt changes in rain intensity. Change in rate is called *rain gradient.* The greater or steeper the gradient, the greater the turbulence.

During a *Callback* conversation the reporter restated the fact that the turbulence was totally unexpected. The pilot stated that the color radar that was on board did not paint the smaller cells and that might have had a bearing on the incident. Unfortunately, the reporter did not state if the 25 miles clearance was from the center of the cell, the level 2 or 3 contour, or the edge of the precipitation. It appears it is possible the airplane may have been well within 20 miles of level 1 returns from the echo, which would indicate the probability of turbulence.

CASE STUDY

Rain showers were in the area. I received a heading from ATC for weather avoidance and was advised a previous aircraft had flown the prescribed route without incident. We had no excessive radar return and no contouring on the radar scope. Flight conditions were IFR, light rain, intermittent light chop. About 70 miles from destination we began experiencing excessive banking and pitching, which I reported to ATC as severe turbulence. I requested immediate deviation and descent as the aircraft was barely controllable. This was approved. The intensity of the turbulence remained severe intermittently for a period of about 7 minutes. During this period we reversed direction of flight and landed at an alternate.

The reporter went on to say that this incident occurred because aircraft and center radars are unable to detect wind direction changes associated with this type of turbulence. A possible solution would be Doppler radar. Certainly Doppler radar can detect changes in wind direction and intensity, but only if precipitation is present. Doppler radar is unable to directly detect turbulence and is of no use in clear air.

The reporter concluded that this encounter was unavoidable since we were unable to detect it on radar. Pilots must remember the limitations on radar. From the reporter's account, this was a rapidly building cell, with limited precipitation in the updrafts. If you fly in the vicinity of convective activity, be prepared for a possible encounter with severe turbulence.

One technique to minimize a pilot's exposure to thunderstorm hazards is to approach a line of convective activity at a 45° angle. This is the same technique used by low-altitude VFR pilots when approaching a mountain pass. If the thunderstorm area does not allow a safe corridor between cells, the pilot has only to make a 90° turn to effect a retreat.

The clear air on the inflow side (upwind) of a storm is a place where severe turbulence can occur. Severe turbulence should be anticipated up to 20 nm from the radar edges of severe storms. These storms often have well-defined radar echo boundaries. This distance decreases to approximately 10 nm with weaker storms that can sometimes have indefinite radar echo boundaries. The wind extending downwind of a storm produces additional turbulence—similar to mechanical turbulence caused by air currents passing around a mountain peak. The rule of thumb is to allow 1 nm for each 1 knot of wind speed.

The rule of thumb is 40 miles between severe storms. The problem is that it may be next to impossible to determine the severity of a storm, especially during the development stage. As with the general

CASE STUDY

Along our route we would encounter numerous thunderstorms and lines of cumulonimbus. We successfully navigated through the lines of cumulonimbus. We could see cumulonimbus south of us and north of our course on radar. Our radar showed several strong echoes over the arrival fix. We informed ATC and they said they could not see them. We asked for a ride report, and ATC said they had no aircraft in that area for awhile. Our radar showed a clear area south of our present position. We asked for and received clearance to deviate and then pickup the rest of the arrival.

I was flying the aircraft on autopilot. I initiated a right turn. We were in the clouds and experiencing a relatively smooth flight. As the turn continued we temporarily broke out of the clouds and saw a rapidly developing cumulonimbus cloud in front of us. There was no indication of this on radar. We had excellent returns on the radar throughout the flight to this point. When I saw the cloud I immediately disengaged the autopilot and increased the angle of bank to avoid it. We penetrated the side of the cloud and experienced severe turbulence for about 4 to 5 seconds. The aircraft climbed rapidly, even though I applied full forward stick. This was a brief encounter.

CASE STUDY

While enroute we descended from FL310 to FL220 for clouds. Deviating north of course, we rounded the corner to proceed direct. Radar showed one more area of thunderstorms for us to navigate around. Radar and visual cues showed an approximately 30-mile gap between storms with sunshine between. As we approach this area we again received clearance to deviate. As we passed north of the first area we turned left, the clearest direction with nothing on radar. As we finished our turn we were in and out of cirrus. At 12 o'clock, less than $1/2$ mile away was a small thunderstorm with tops estimated to FL230. With no room to turn, the captain put the ignition to override just prior to entering. The aircraft started to climb. The autopilot could not keep up and kicked off about the time of our one big jolt of turbulence. The captain was following on the controls and in a few seconds we popped out in smooth air 200 to 300 ft above assigned altitude.

CASE STUDY

MSY UUA /OV NEW 150020/TM 2015/FLDURD/TP C550/TB SEV/RM OCCURED IN AREA WHERE ACFT RADAR DID NOT INDICATE PCPN. BOTH CREW INJURED.

limitations on radar, if we fudge on the 40-mile rule, we should be prepared for a significant turbulence encounter.

This incident occurred over New Orleans in thunderstorm weather. Both crew of a Cessna Citation were injured when the aircraft encountered severe turbulence in an area where their airborne weather radar indicated no precipitation.

Pilots should avoid takeoff or landing when a thunderstorm is within 10 to 20 miles of the airport. This is the region of the strongest and most variable winds. Caution must also be exercised following thunderstorm passage. A strong, gusty outflow boundary can follow the storm. Along with these winds are downbursts and microbursts that produce severe low-level wind shear.

Note that in addition to the gust, wind was variable between 210° and 030°, this resulting in a 180° variation in wind direction, a classic indication of a microburst.

CASE STUDY

American Airlines flight 1420 crashed while attempting to land at Little Rock's National Airport. The crash occurred just before midnight on June 1, 1999. On approach a microburst hit the airport with a wind gust to 76 knots.

KLIT 020458Z 29010G76KT 210V030 1/2SM + TSGSRA...

The FAA, along with a group of aviation specialists, has developed AC 00-54, *Pilot Windshear Guide.* Although primarily for the airlines, much of the information can be applied to general aviation. The Department of Commerce publication *Microbursts: A Handbook for Visual Identification* is for sale by the Superintendent of Documents. It contains an in-depth, technical explanation of the phenomena along with numerous color photographs depicting microburst activity. Avoidance is the best defense against a microburst encounter.

The first technique in thunderstorm avoidance is a complete preflight weather briefing. "Forewarned is forearmed." The second technique is the venerable "see and avoid" concept. Pilots should look for the tell-tale signs of convective activity. The following is a list of visual convective turbulence, lightning, hail, downburst, microburst, and severe wind shear indicators.

- Anvil cloud form approaching.

- Darkened color to clouds.

- Churning vertical clouds.

- Vertical clouds that are growing.

The third step in weather avoidance is the use of weather radar.

Depending on the type and movement of precipitation, pilots need to be able to recognize hazardous thunderstorm patterns. Recommended thunderstorm avoidance practices are listed below. Avoid by at least 20 nm the following conditions:

- Any storm with tops at or greater than 20,000 ft.

- All rapidly moving echoes.

- The entire cell, if any portion of the cell contains level 3 returns.

- All rapidly growing storms.

- All storms showing erratic motion.

- Never continue flight toward or into an area where radar attenuation is occurring.

• Steep rain gradients.

• Takeoff and landing within 10 to 20 miles of a storm.

One key to hazardous weather avoidance is severe echo recognition. The following discussion describes the echo patterns that are typical of severe weather and are illustrated in Fig. 1-7.

Probably the most familiar—certainly the most discussed—is the hook echo. Hooks are typically located at the right rear side of the thunderstorm's direction of movement, most often the southwest quadrant. The hook is a small-scale low-pressure area, sometimes referred to as a *mesocyclone* or *mesolow*. It typically ranges from about 3 to 10 miles in diameter. Tornadoes (FC) form in the low region near the hook. Research indicates almost 60 percent of all observed hook echoes spawn tornadoes. However, a hook echo does not always generate a tornado.

Finger echoes, like a hook, represent a strong probability of severe thunderstorms and tornadoes.

Bow-shaped echoes are typically associated with fast-moving, broken or solid lines of thunderstorms. Severe weather will most likely develop along the bulge and at the northern end of the echo pattern. Like hook echoes, bow echoes may spawn tornadoes, as well as strong, gusty surface winds and hail.

Large, isolated echoes will sometimes have the configuration of a V or U shape—a V notch. A V notch is often accompanied by severe thunderstorms and tornadoes. The severe weather systems, including tornadoes (FC), are most often located on the southwest quadrant of the storm, like the hook echo. But, here again, severe weather does not ecessarily accompany a V notch.

Line echo wave patterns (LEWPs) are caused by one portion of a line accelerating, causing a wavelike configuration. LEWPs

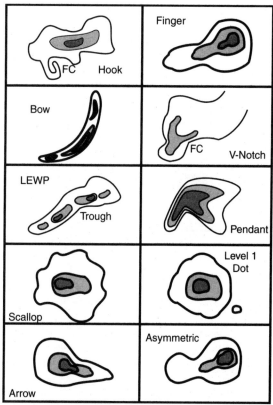

Fig. 1-7. A key to hazardous weather avoidance is severe echo recognition.

form solid or nearly solid lines. Typically the most severe weather occurs in the trough of the wave. Additional areas of severe weather are associated with the advancing edge of the line (east or southeast).

A pendant represents one of the most severe storms—a supercell. Supercells produce maximum average hail over 2 inches in diameter, with hail swaths over 12 miles, and 60 percent produce funnel clouds or tornadoes.

Scalloped-edge echoes indicate turbulent motion within the cloud. There is a good probability of hail associated with these echoes.

Hail is a poor radar reflector and often falls outside the storm. Expect hail to fall downwind of the parent cell. This may be indicated on radar as an isolated level 1 dot not attached to the convective storm.

Severe storms are tilted through the atmosphere. This is what allows them to be steady state and often severe. This tilting is sometimes indicated on radar by an arrow shape. Like the arrow shape, any asymmetric echo indicates a tilted storm, with its associated severe weather. The storm produces echo shapes and colors that are not even or concentric.

Figure 1-8 depicts recommended thunderstorm avoidance procedures. The dotted line surrounding the storm represents the significant turbulence envelope. It extends 20 miles laterally from the cloud for severe storms. Pilots should avoid flight above tops by 1000 ft for each 10 knots of wind at storm top. Avoid the entire cell if any portion contains level 3 echoes or greater. To avoid hail, do not fly under the anvil cloud. Pilots should avoid takeoff and landing within 10 to 20 miles of a storm.

Note the airborne radar display in Fig. 1-8. The cloud boundary is much closer to the airplane than the radar return. The radar is indicating level 3 or greater returns from the center of the storm. The radar indicates the weakest part of the storm almost directly ahead. However, there is a dark area behind the storm, with an absence of echoes. This is a classic indication of radar attenuation. The pilot should definitely not take this course. In fact, the entire storm should be avoided. Also note the single level 1 return, adjacent to the right side of the storm as seen on the radar display. This is an indication of

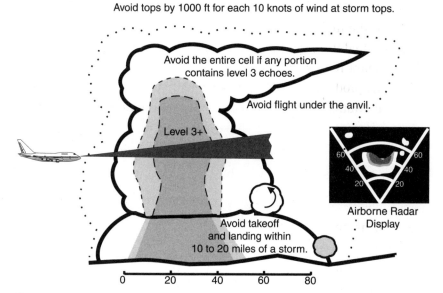

Fig. 1-8. Avoidance is the key to the thunderstorm threat.

hail. Therefore, on the basis of the information available, the pilot should alter course to the left side of the storm, as seen on the radar.

The following are some dos and don'ts of thunderstorm avoidance:

- Do avoid by at least 20 miles any thunderstorm identified as severe or giving an intense radar echo; this is especially true under the anvil of a large cumulonimbus.

- Do clear the top of a severe thunderstorm by at least 1000 ft for each 10 knots of wind speed at the cloud top.

- Do regard as severe any thunderstorm with tops 35,000 ft or higher.

- Don't land or take off in the face of an approaching thunderstorm.

- Don't attempt to fly under a thunderstorm, even if you can see through to the other side.

- Don't try to circumnavigate thunderstorms covering more than half of the area, even with storm detection equipment.

- Don't attempt to enter areas of embedded thunderstorms without storm detection equipment.

If thunderstorm penetration cannot avoided, the following steps are recommended before entering the storm:

- Tighten seat belts and shoulder harnesses, secure all loose objects.

- Plan a course through the storm in the minimum of time.

- To avoid the most critical icing, establish a penetration altitude below the freezing level or above $-15°$ C.

- Turn on pitot heat and carburetor or inlet heat.
  Establish power setting for turbulence penetration airspeed.

- Turn up cockpit lights to highest intensity to lessen danger of temporary blindness from lightning.

- Disengage autopilot altitude and speed hold.

- Tilt airborne radar antenna up and down occasionally. Tilting may help detect a hail shaft or a growing thunderstorm cell.

If all of your best efforts fail, the following are some dos and don'ts to observe during thunderstorm penetration:

- Do keep your eyes on the instruments. Looking outside increases the danger of lightning blindness.

- Do maintain a constant attitude; let the aircraft ride with the turbulence. Maneuvers to maintain altitude increase gust loading.

- Don't change power settings.

- Don't turn back once in the thunderstorm. A straight course through the storm most likely will get you out of the hazards most quickly. Turning increases gust loading.

These are recommendations and, as we've seen, do not necessarily mean significant turbulence and hail will always accompany these parameters. Pilots do encroach on these limitations with no significant adverse affects. However, NASA Aviation Safety Reporting System reports confirm the existence of hazardous weather within these limitations. It's the pilot's decision to accept or reject these margins.

Three final words regarding thunderstorms: avoid, avoid, avoid!

CHAPTER 2

Thunderstorm Avoidance Systems

Thunderstorm avoidance systems fall into three categories: radar, lightning detection systems, and the "Mark 1 Eyeball"—visual avoidance. Radar and lightning detection systems allow the pilot to avoid hazards without direct visual contact. However, both radar and lightning detection systems have limitations.

Pilots have access to National Weather Service (NWS) radar—through the Internet, FAA Flight Service Stations, and commercial vendors—to some extent ATC radar, and airborne weather radar. With the exception of ATC radar and airborne weather radar, radar images are on the order of 5 to 15 minutes old. This time delay also applies to ground-based data uploads from commercial vendors in cockpit weather displays. For most practical purposes the products discussed in this chapter are "real time." That is, real time as opposed to the graphic and textual products discussed in Chap. 3, "Radar Weather Products."

Since thunderstorms can develop at an astonishing rate, to safely negotiate convective activity pilots must have access to real-time information in flight. This means airborne weather radar, lightning detection equipment, or both, flight deck access to radar information, contact with Flight Watch, or visual sighting to evade individual storms. Although ATC radars, to a limited extent, display weather, pilots

> **NOTE**
>
> Because of the astonishing rate at which thunderstorms can develop, building at up to 7000 ft/min and moving over the ground at 35 to 45 knots, it can be argued that the only real "real time" products are ATC and airborne weather radar, or airborne lightning detection systems. For most practical purposes, time frames on the order of several minutes are adequate. However, like anything else in the realm of aviation weather, there are no absolutes or guarantees.

cannot depend on controllers to keep them out of harm's way—unless the pilot declares an emergency. We'll discuss this issue further in subsequent chapters.

The FAA is in the process of providing Air Traffic Controllers with either separate or overlay NEXRAD products. This will assist controllers in providing pilots with additional weather avoidance. However, as we shall see, these services are not a substitute for airborne weather avoidance equipment.

Each system, ground-based or airborne, has a specific purpose and its own applications and limitations. Many of these were discussed in Chap. 1. For a pilot to safely apply these tools, their purpose and limitations must be clearly understood.

The final section of the chapter contains a discussion of the National Convective Weather Forecast (NCWF). The NCWF is a new product, using real-time radar and lightning observations to extrapolate significant convection over the next hour. It is not a panacea for convective weather forecasting, but an additional tool to help pilots avoid hazardous weather. At present it is available only on the Internet and, like every other product or forecast, has limitations that must be understood.

National Weather Service Radars

NWS radars, with a narrow linearly polarized beam, are ideal for detecting precipitation-size particles. Sensitivity time control on NWS radars compensates for range attenuation—the loss of power density due to distance from echoes. STC-displayed intensity remains independent of range; therefore, targets with the same intensity, at different ranges, appear the same on the display. NWS radars can detect targets up to 250 nm; however, because of limitations in range and beam resolution (the ability of the radar to distinguish individual targets at different ranges and azimuths), an effective range of 125 nm is used.

NEXRAD has been a quantum leap in providing early warning of severe weather. NEXRAD will be the standard for the next 20 to 25 years, with a wavelength of 10 cm. The WSR-88D network will consist of up to 195 units, 113 NWS sites in the contiguous United States, with additional sites in Alaska, Hawaii, the Caribbean, and western Europe, and 22 Department of Defense (DOD) units. The NEXRAD network fills most of the radar gaps, previously without weather radar coverage, in the western United States. NWS weather radar locations are contained in Appendix B.

Since NEXRAD is a Doppler radar, it detects the relative velocity of precipitation within a storm. It has increased the accuracy of severe thunderstorm and tornado warnings, and has the capability of detecting wind shear. Figure 2-1 shows a NEXRAD weather radar site. The radar specialist using a WSR-88D will have Radar Data Acquisition, Radar Product Generation, and Display units. NEXRAD provides hazardous and routine weather radar data typically above 6000 ft east of the Rockies and above 10,000 ft in the western United States. It can also be made available in the aircraft through Mode S transponders or other systems with automatic data link capability.

Of the NEXRAD WSR-88D radar products, currently four are available through the National Weather Service's Internet site at weather.noaa.gov/radar/national.html. Most local NWS offices and the Aviation Weather Center have links to this site. Products currently available are

- Base reflectivity

- Composite reflectivity

- One-hour precipitation

- Storm total precipitation

Fig. 2-1. National Weather Service and military NEXRAD radars cover most of the continental United States.

Of primary interest to aviation are base reflectivity and composite reflectivity. The availability of a "loop" of images has certain advantages. For example, the loop can be used to determine echo movement (direction and speed), echo intensity trend (increasing or decreasing), and to help detect nonprecipitation echoes such as ground clutter and anomalous propagation.

Base reflectivity images are available at several elevation angles (tilts) of the antenna and are used to detect precipitation, evaluate storm structure, locate atmospheric boundaries, and determine hail potential. The base reflectivity image currently available from the NWS Internet site is the lowest "tilt" angle (0.5°). This means the radar's antenna is tilted 0.5° above the horizon.

The maximum range of the base reflectivity product is 124 nm from the radar location. This view will not display echoes that are more distant than 124 nm, even though precipitation may be occurring at greater distances. To determine if precipitation is occurring at greater distances, an adjacent radar or link to the National Reflectivity Mosaic must be used.

Composite reflectivity displays maximum echo intensity from any elevation angle at every range from the radar. This product is used to reveal the highest reflectivity in all echoes. When compared with base reflectivity, the composite reflectivity often reveals important storm structure features and intensity trends.

The maximum range of the composite reflectivity product is 248 nm from the radar site. The blocky appearance of this product is due to its lower spatial resolution of a 2.2 by 2.2-nm grid. It has one-fourth the resolution of base reflectivity.

Although the composite reflectivity product is able to display maximum echo intensities as far as 248 nm from the radar, the beam of the radar at this distance is at a very high altitude in the atmosphere. Thus, only the most intense convective storms and tropical systems will be detected at the longer distances. Because of this fact, special care must be taken in interpreting this product. While the radar image may not indicate precipitation, it's quite possible that the radar beam is overshooting precipitation at lower levels, especially at greater distances. To determine if precipitation is occurring at greater distances, an adjacent radar site or the National Reflectivity Mosaic should be used.

Figure 2-2 illustrates the difference between base and composite reflectivity, and the short-range and long-range images available. (The short-range base reflectivity image is on the left of Fig. 2-2; the long-range, composite image is on the right of Fig. 2-2.) The short range provides more detail than is available with the long-range image. Note

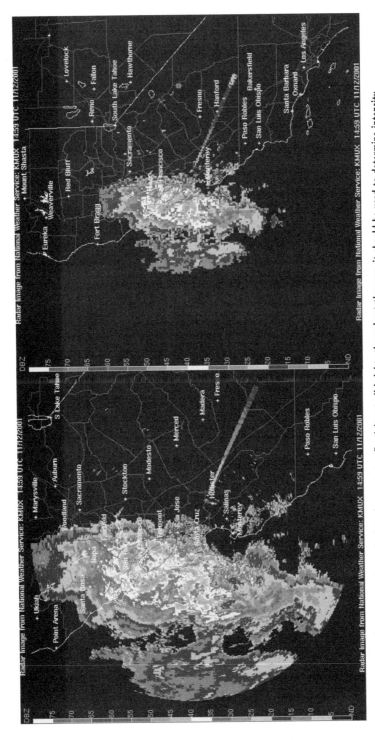

Fig. 2-2. Because the composite image depicts the greatest reflectivity at all heights throughout the scan, it should be used to determine intensity.

the difference in the image scales. Because the composite image depicts the greatest reflectivity at all heights throughout the scan, it should be used to determine maximum intensity. The composite image shows more intense precipitation in the center of this weather event. Notice the steep rain gradient on the east side of the area of precipitation, as compared with the north, west, and south. The radial spoke extending well to the southeast is an example of anomalous propagation (AP).

Radar products are available in precipitation mode and clear-air mode. Precipitation mode is designed to primarily detect precipitation-size objects. In clear-air mode, the radar is in its most sensitive operation. This mode has the slowest antenna rotation rate, which permits the radar to sample a given volume of the atmosphere longer. This increased sampling increases the radar's sensitivity and ability to detect smaller objects than is possible in precipitation mode. Clear-air mode detects airborne:

- Blowing dust

- Updrafts

- Insect swarms

- Cloud street

- Smoke plumes

- Bird migrations

- Inversion layers

- Fronts

- Sea breezes

- Outflow boundaries

Never assume that in clear-air mode there are no precipitation echoes. However, most often these returns will be nonprecipitation, but the high end (above about 16 dBZ) may result from light rain or snow. Recall that snow is a poor reflector of radar energy. Therefore, clear-air mode will occasionally be used to detect light snow. In the winter, dBZ values between 4 and 16 (yellow) would represent light snow. Snow flurries will appear as a 4-dBZ value. Snow characteristics—fine or large, wet, or dry—all play a role in how intense the dBZ levels will appear. An area of dry, large snow flakes may appear only between 8 and 12 dBZ.

Precipitation and clear-air modes are most often clearly distinguishable. The key is the dBZ scale. In precipitation mode, the scale goes from +5 to +75 dBZ; in the clear-air mode, the scale goes from −28 to +28 dBZ. This is illustrated in Fig. 2-3. Note the differences in the dBZ scales.

In addition to the radar limitations already discussed, radar coverage is limited close to the site by its inability to scan directly overhead. Therefore, close to the radar site, data is not available because of the radar's maximum tilt elevation of 19.5°. This area is commonly referred to as the radar's "cone of silence." To determine the activity directly over a site, an adjacent radar should be used.

An additional limitation affects some radar sites in the west. A number of sites are located on mountain tops, thus restricting coverage to above 10,000 ft. One result is that valley surface observations may report precipitation that is not being detected on radar. However, this precipitation is usually light and not the result of convective activity.

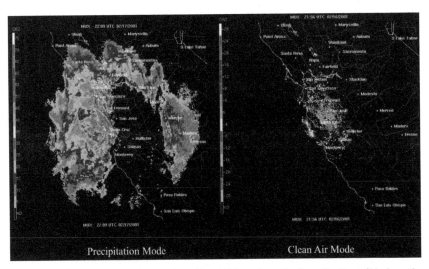

Fig. 2-3. The difference between precipitation and clear-air modes is easily discernible from the dBZ scale.

Image updates are based on the operation mode of the radar at the time the image is generated. As previously mentioned, the WSR-88D Doppler radar is operated in precipitation or clear-air mode. In precipitation mode, images are updated every 5 or 6 minutes. In clear-air mode, images are updated every 10 minutes.

Echoes from surface targets appear in almost all radar reflectivity images. In the immediate area of the radar, "ground clutter" generally appears within a radius of 20 nm. This appears as a roughly circular region with echoes that show little spatial continuity. It results from radio energy reflected back to the radar from outside the central radar beam, from Earth's surface, or from buildings.

Under highly stable atmospheric conditions (typically on calm, clear nights), the radar beam can be refracted almost directly into the ground at some distance from the radar, resulting in an area of intense-looking echoes. This "anomalous propagation" (AP) phenomenon is much less common than ground clutter. Certain sites situated at low elevations on coastlines regularly detect "sea return," a phenomenon similar to ground clutter except that the echoes come from ocean waves. AP often appears as a "spoke" on the radar that is unrelated to other weather returns. An example of AP is shown in Fig. 2-2.

Returns from aerial targets are also rather common. Echoes from migrating birds regularly appear during nighttime hours between late February and late May, and again from August through early November. Returns from insects sometimes appear during the months of July and August. The apparent intensity and areal coverage of these features are partly dependent on radio propagation conditions, but they usually appear within 30 nm of the radar site and produce reflectivity of less than 30 dBZ.

Flight Service Stations (FSSs) and Center Weather Service Units (CWSUs) have access to real-time NEXRAD weather radar products. FSS controllers are specially trained to interpret information for preflight briefings and inflight weather updates. FSS and Flight Watch controllers translate radar echo coverage into the following categories:

• Widely scattered: less than 1/10

• Scattered: 1/10 to 5/10

- Broken: 6/10 to 9/10

- Solid: More than 9/10

Limitations exist. West of the Rockies, sites suffer from ground clutter, although the implementation of the NEXRAD network has helped to eliminate ground clutter to a large degree.

FSS and Flight Watch controllers have access to a number of radar displays. Figure 2-4 illustrates a National Weather Radar Summary product. This display is useful to obtain an overall view of precipitation and convective activity over the contiguous United States. It displays intensity, movement, and tops of echoes. It may also contain information on severe weather hazards, such as mesocyclones or mesolows—these are tornadic signatures and may indicate large hail. Like the National Reflectivity Mosaic and the Radar Summary Chart (both discussed in Chap. 3), this product is derived from individual Radar Weather Reports (RAREPS), also discussed in Chap. 3. Because of the time delay from observation, through coding and dissemination, to product presentation, this display is often old by the time it

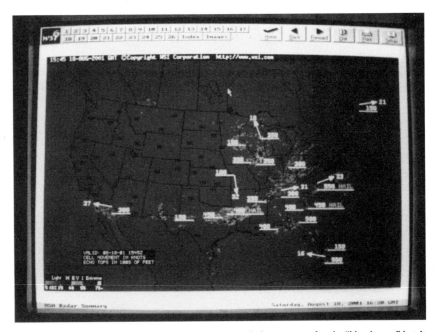

Fig. 2-4. The National Weather Radar Summary is a good place to start for the "big picture," but it cannot be used for weather avoidance.

becomes available. The same caution holds true for national weather radar presentations available through commercial vendors and on TV. It's a good place to start for the "big picture," but cannot be used for inflight weather avoidance.

In addition to a National Weather Radar Summary, FSS controllers have access to regional, local, and individual radar site presentations. Regional and local composites assist controllers in relaying information on the intensity, location, and movement of precipitation echoes. Often a "loop" is available to assist in the determination of weather movement and echo intensity trends. One such composite is illustrated in Fig. 2-5, which shows northern Illinois and parts of the adjacent states. In addition to providing the location and intensity of precipitation echoes, these presentations are useful in determining the safest and most advantageous flight routes to allow pilots to avoid hazardous weather. For example, the FSS or Flight Watch controller can advise the pilot of the extent of convective weather, where weather is developing or dissipating, and the best direction to escape hazardous weather.

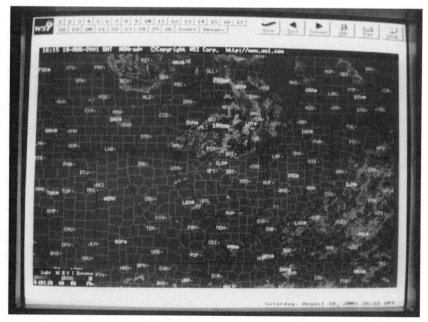

Fig. 2-5. FSS or Flight Watch controllers can advise pilots of the extent of convective weather, where weather is developing or dissipating, and the best direction to escape hazardous weather.

Figure 2-6 illustrates a local weather radar composite display for the Chicago area. These composites are useful to advise pilots of the specific radar picture. For example, a pilot approaching the Chicago area from the northwest would be well advised to land short, rather than attempting to penetrate the weather northwest of the destination. Or, a pilot on a southeast-bound course would be advised of the extent and southern boundary of the convective activity, in order to alter course prior to encountering hazardous weather. Even pilots with airborne weather avoidance equipment can use these services to evaluate the extent of a convective weather system.

The Flight Service Stations also have access to individual weather radar sites. These displays are similar to those contained in Figs. 2-2 and 2-3, with similar advantages and limitations.

Private vendors also have access to radar data. Information from radars is coded and transmitted. When you use radar information, it's important to know how individual vendors display information such as intensity and if it is indeed a real-time observation, or a freeze or memory display. The display may be in the precipitation or clear-air mode.

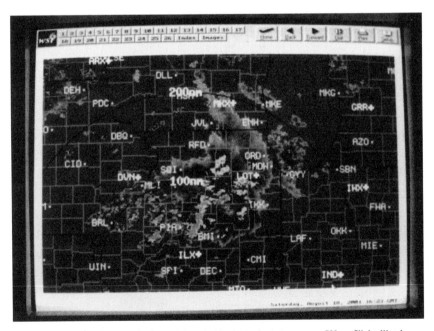

Fig. 2-6. Even pilots with airborne weather avoidance equipment can use FSS or Flight Watch services to evaluate the extent of a convective weather system.

Air Traffic Control Radar

ATC radar is specifically designed to detect aircraft; a narrow fan-shaped beam reaches from near the surface to high altitudes. ATC radars have a wavelength of 23 cm, which is ideal for detecting aircraft, but reduces the intensity of detected precipitation; additional features reduce the radar's effectiveness to see weather.

To efficiently detect aircraft, and eliminate distracting targets, ATC controllers activate circular polarization (CP), moving target indicator (MTI), and sensitivity time control (STC) circuits.

CP results in a low sensitivity to light and moderate precipitation. MTI displays only moving targets; unless droplets have a rapid horizontal movement they remain undetected; even rapidly moving precipitation will not be observed when advancing perpendicular (tangentially) to the radar beam. STC further eliminates light precipitation and decreases the intensities of displayed precipitation. Naturally, controllers, especially at approach control facilities, engage these features during poor weather to accomplish their primary task: aircraft separation.

Air Route Traffic Control Center (ARTCC) radars measure the signal strength of radar returns from precipitation to obtain an indication of storm intensity. It does not display areas of light precipitation considered operationally insignificant. ARTCC radars display precipitation returns in two modes. Radial lines indicate lower-intensity precipitation. The letter H represents areas of higher-intensity precipitation, which might be thunderstorms. This is illustrated in Fig. 2-7, a composite radar display from several ARTCC radar sites. Note the radial lines emanating from the radar antenna. Also note the "cone of silence" that is void of lines in the center of the illustration. The system is unable to detect precipitation within an 8-nm radius of the radar site. Weather returns can be omitted or distorted over the antenna. Because of antenna tilt, weather radar returns are restricted to 200 nm from the antenna and limited weather data is displayed below FL350 and 200 nm. In the center, right of the figure are a number of H symbols indicating heavier precipitation. Figure 2-7 also illustrates the clutter that occurs when precipitation is displayed.

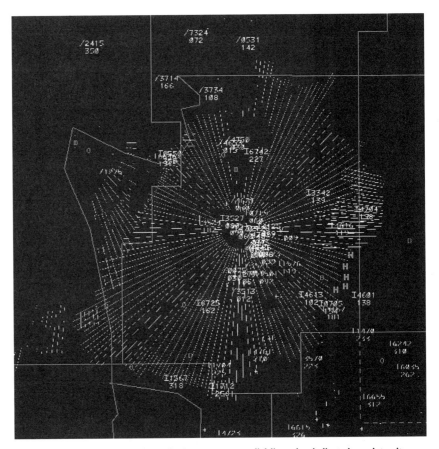

Fig. 2-7. ARTCC radars display precipitation returns as radial lines that indicate lower-intensity precipitation; the letter H represents areas of higher-density precipitation, which might be thunderstorms.

The controller has the option of selecting the Weather Fixed Map Unit (WFMU) with the weather 1 key (WX 1; lines level 1 and 2 echoes) and the weather 2 key (WX 2; H level 3 and greater echoes). When the precipitation intensity just barely qualifies for the H symbol, the precipitation can be moderate. Since the precipitation intensity may greatly exceed the minimum H parameter, the storm may be severe and should be treated as the worst case.

Controllers are instructed to select WX 1 and WX 2 in accordance with weather and operational requirements. (Recall the clutter in Fig. 2-7. Especially in areas of high-density traffic, the controller is not going to be inclined to use this feature.) The controller will select both keys when providing radar inflight weather avoidance assistance. Pilots must understand that the area of precipitation displayed to the controller may represent only part of the storm. Because of the limitations on ATC radars, the pilot's view of the weather as seen on airborne weather radar can be significantly different from that observed by ATC personnel. Studies also show considerable difference between ATC radar's location and intensity of precipitation as compared to NWS NEXRAD Doppler radars. The differences are due to the limitation of ATC radars.

Early Airport Surveillance Radars (ASR) used various internal circuits to minimize weather returns. This was especially necessary when controllers were required to separate nontransponder, primary targets. These circuits included circular polarization (CP), moving target indicator (MTI), and sensitivity time control (STC).

The ASR-9 has the capability to detect all six levels of precipitation. However, the controller can display only two levels at one time on the radar presentation—low (dim) and high (bright). For example, let's say that levels 1, 2, and 3 are being detected by the ASR-9. If the controller selects 1, all returns, whether 1, 2, or 3, would appear at the same intensity on the display: low (dim). Should the controller select 3 as well as 1, level 3 returns would be displayed as high (bright), with levels 1 and 2 as low (dim).

The amount of weather avoidance assistance that a controller can provide depends on a number of factors. What type of equipment is available to the controller? How much training and experience does the controller have in using the weather detection circuits? What is the traffic activity within the controller's area of responsibility?

The FAA is in the process of upgrading weather depiction for air traffic controllers. One such initiative is the Weather and Radar Processor (WARP). WARP generates a NEXRAD mosaic. Other systems depict NEXRAD images on the controller's displays. Both systems add clutter to the display, which is of paramount concern to the controller, whose primary task is to maintain the separation of aircraft.

The following case study illustrates the limitations of ATC radar and a pilot's overreliance on controller-provided weather avoidance assistance. The study came from a NASA Aviation Safety Report System report.

CASE STUDY

On an IFR flight I was switched from approach to center after being cleared to deviate west of course. I could hear center talking to other aircraft, but they failed to respond to my check-in for some 6 to 8 minutes. During this time I entered heavy weather that did not show on the passive (lightning detection) weather-indicating system. Weather included heavy rain and severe turbulence. After autopilot disconnect and slowing of airspeed, I was unable to hold altitude. Center finally responded to my calls. They told me they could not help me and switched me to approach control. After about a minute with approach, I broke into the clear.

I realize not all center radar has weather graphic overlay. If center had been able to see the weather in my flight path and if they had not been so overloaded that they could not talk to me, I would have had a safer, more comfortable flight. I also realize the limitations of the passive weather-indicating system equipment.

From the pilot's statement it appears the aircraft was within congested airspace. Recall that ATC's primary responsibility is the separation of known aircraft. When controllers are very busy separating aircraft, they will tend to ignore calls from aircraft that do not require immediate attention for separation purposes. (Controllers are instructed to select WX 1 and WX 2 in accordance with weather and operational requirements.) This appears to be the issue in this case. The pilot had an overexpectation of ATC capabilities. The pilot indicated that the lightning detection system on board did not show any significant activity. (We'll discuss the limitations of lightning detection equipment in a subsequent section.) Pilots can never put blind faith in ATC to keep them out of convective activity.

Airborne Weather Radars

When using an airborne weather radar it is imperative to understand the particular unit, its operational characteristics, and limitations. "Just reading through the brochure that comes with the equipment is certainly not enough to prepare a pilot to translate the complex symbology presented on the [airborne] scope into reliable data. A training course with appropriate instructors and simulators is strongly recommended," according to the March–April 1987 *FAA Aviation News*.

Airborne weather radars are low power, generally with a wavelength of 3 centimeters. Precipitation attenuation, which is directly related to wavelength and power, can be a significant factor. An accumulation of ice on the aircraft's radome causes additional distortion. Maintenance of the radome is another issue. Damaged or improperly repaired radomes can render units all but useless. Figure 2-8 shows the radome of a Piper Seneca.

In Chap. 1, the limitations on the size of the radar's antenna were discussed. For most general aviation airplanes, a 10-inch antenna is often the largest practical size. Figure 2-8 shows a typical radome installation for a multiengine general aviation airplane. Single-engine airplanes present a different problem. Since the nose of the airplane contains the engine, the radome must be placed in or under the wing. To accomplish this some manufacturers attach a pod underneath the wing. Others place the radar antenna in the wing, as illustrated in Fig. 2-9.

Fig. 2-8. Damaged or improperly repaired radomes can render units all but useless.

Fig. 2-9. For single-engine airplanes, the radar antenna must be placed in or under the wing.

Most air carrier airplanes, like the one shown in Fig. 2-10, can accommodate larger antennas. This airplane can be equipped with a 30-inch antenna. With antennas, bigger is better.

Airborne weather radar display units are usually about the same size. A typical general aviation display unit is illustrated in Fig. 2-11. However, the degree of detail and the usable range of the unit remains dependent on the size of the antenna. Some manufacturers combine the radar display with other avionics features. For example, the weather radar display may be superimposed on a navigational display. Like ATC controllers, pilots need to be careful not to put so much information on one display that needed information becomes obscured.

In September 2001, an FAA User-Needs Analysis Group met in Washington, D.C. The issue was thunderstorm intensity levels and descriptions used by air traffic controllers and pilots.

Prior to the implementation of NEXRAD, the NWS and FAA used the six intensity levels described in Table 1-2. Although NEXRAD displays a greatly increased number of levels, for aviation purposes

Fig. 2-10. With antennas, bigger is better; most air carrier airplanes can accommodate a 30-inch antenna.

Fig. 2-11. The degree of detail and the usable range of the weather radar depend on the size of the antenna.

TABLE 2-1. Proposed Pilot/Controller Radar Intensity Levels

PRECIPITATION INTENSITY	dBZ	NEW DESCRIPTOR
1	20–30	Light
2	30–40	Moderate
3	40–45	Heavy
4	45–50	Heavy 4
5	50–55	Heavy 5
6	55+	Heavy 6

the same six levels of intensity have been retained. However, airborne weather radars display only three or four levels. The Group agreed and created the scale in Table 2-1.

The *light, moderate,* and *heavy* terms in the table correspond directly to current green, yellow, red, and purple contours from aircraft radar using the "automatic gain" feature. At present the descriptors are only a proposal, but the table very effectively illustrates the difference between ground-based and airborne weather radar intensity level.

Lightning Detection Systems

Meteorological conditions in thunderstorms produce lightning—the visible electrical discharge produced by a thunderstorm. (Thunder is the sound generated by the rapidly expanding gases along the channel of a lightning discharge.) Typically, the greater the amount of rain and

NOTES

For a weather observer to report a "thunderstorm," thunder must be heard, or if ambient noise levels are high, lightning must be seen. The Automated Lightning Detection and Reporting System (ALDARS), which acquires lightning information from the National Lightning Detection Network, will allow automated observation sites to report the occurrence of a thunderstorm.

The 3/4-inch hail criteria for a severe thunderstorm was established in 1954; the wind criteria was lowered in 1970 to 50 knots for aviation purposes. Since 1996 there is no longer a METAR weather code to alert pilots to the occurrence of a severe thunderstorm, but the clues appear in the wind and weather phenomena groups, and remarks of the report. There is a proposal to increase severe thunderstorm criteria to winds of 52 knots (60 mph) and hail to 1 inch for public use. This would substantially change the number of warnings issued, the goal being to reduce overwarning. However, as we have seen, the threshold for damaging hail to aircraft is 3/4 inch.

hail, the greater the production of lightning. For a storm to be labeled a thunderstorm, thunder and lightning must be present.

In a storm cell, updrafts and downdrafts create areas of different electrical potential. One theory proposes that a typical cloud has three areas of charge. These are illustrated in Fig. 2-12. The main area of negative charge occurs at around 19,000 ft, and is approximately 3000 ft thick. Higher in the cloud is a positive area extending well above the freezing level. Another area of negative charge exists in the very light ice crystals at the top of the cloud. Studies show the most active area is at temperatures between $-6°C$ and $+11°C$, with some type of precipitation. However, lightning may occur at any level within the thunderstorm or even outside the thunderstorm itself. Lightning can occur in the clear air around the top, sides, and bottom of a storm or in the form of a "bolt out of the blue" several miles from the thunderstorm cell. Studies show that pilots can reduce the risk of a lightning strike by not flying under the following conditions:

- Within $8°C$ of the freezing level
- Within 5000 ft of the freezing level
- In light precipitation, including snow
- In areas of light or no turbulence

Lightning experienced in a thunderstorm can cause temporary blindness so that control of the aircraft by reference to instruments may be momentarily lost. Damage to navigational and electronic equipment can also create a hazard. Direct lightning strokes can cause small punctures in the aircraft skin. Lightning strikes can damage or destroy radomes, burn wire, magnetize airframes, destroy composite structures, fuse control surfaces, cause turbojet compressor stall and flameout, and, very rarely, ignite fuel tanks.

Injuries from a lightning strike can range from none to serious or even fatal. They include various degrees of burns, deafness, and flash blindness. An aircraft, like a car on the ground, typically insulates the passengers from lightning hazards. The only known incident of lightning downing a jetliner occurred on December 8, 1963. Over Elkton, Maryland, lightning struck the aircraft, exploding three of its fuel tanks. Eighty-one people perished. In May 1996 a Beech King Air was struck by

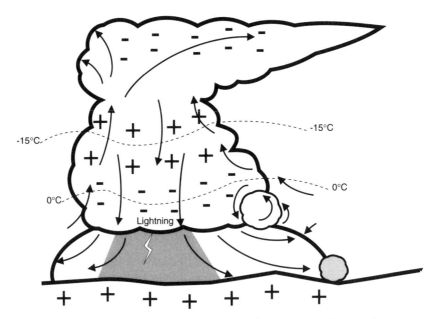

Fig. 2-12. Lightning can occur in the clear air around the top, sides, and bottom of a storm, or several miles from the thunderstorm cell.

lightning, resulting in a cabin fire. Although lightning strikes an airplane approximately every 3000 hours, significant damage is the exception, rather than the rule. The answer: Avoid thunderstorms!

Lightning is classified by its origin and destination. Lightning can be from cloud to ground, in cloud, cloud to cloud, or cloud to air. Astronauts have even observed lightning coming out of the tops of clouds—cloud to space? Figure 2-13 shows lightning that would be classified as cloud to ground.

CASE STUDY

An Air Force commander ordered a military KC135 tanker to penetrate a level 4 thunderstorm to refuel an SR-71. The pilot, who was also a meteorologist, had always wondered about the diameter of a lightning bolt. You guessed it! A lightning stroke punched a 4-inch-diameter hole in the radome. (Allegedly, the commander who ordered the thunderstorm penetration was soon looking for a new position.)

Fig. 2-13. The solution to the lightning hazard is to avoid thunderstorms.

Another electrical phenomenon associated with thunderstorms is corona discharge, colloquially known as *St. Elmo's fire.* St. Elmo's fire becomes visible as bluish static electric streaks dancing across the windscreen. St. Elmo's fire is not a danger in itself. Aircraft flying through or in the vicinity of thunderstorms often develop corona discharge streamers from antennas and propellers, and even from the entire fuselage and wing structure. It produces the so-called precipitation static, or p static. P static, however, usually affects only low-frequency radio communications and navigation. To prevent or reduce p static, aircraft are equipped with static discharge wicks on the trailing edges of the control surfaces. Some pilots have reported precipitation static and St. Elmo's fire as an indicator of a significant charge on the airframe preceding an imminent strike. P-static wicks must not be confused with lightning protection devices, although they may serve as a an exit point for lightning and reduce damage to the aircraft.

Lightning or electromagnetic detectors can display the presence of lightning even before the water content of a storm is sufficient to cause a significant radar return. Since not all level 3 echoes are thunderstorms, an electrical sensor can identify the cell as a thunderstorm through its electrical activity.

Lightning detectors determine range by the strength of the signal of the lightning discharge. Nearby strikes produce a strong signal; distant strikes produce a weaker signal. Distances are derived from the deterioration of electrical strength or deterioration of frequencies over distance. An algorithm determines distance on the basis of a typical discharge. Severe storms appear closer than they actually are, and weaker storms farther than their actual distance. Therefore, distances, especially at the extreme range of the unit, may be inaccurate. Distance accuracy tends to vary by ±15 percent. However, the closer to the aircraft the lightning is, the more accurate the range. Azimuth, on the other hand, is extremely accurate.

Lightning detection systems have limitations. One misconception is that, in the absence of dots or lighted bands, there are no thunderstorms. However, NASA's tests of the Stormscope detection system differed. Precipitation intensity level 3 and occasionally level 4 would be indicated on radar without the lightning detection system being activated. A clear display indicates only the absence of electrical discharges. This does not necessarily mean convective activity and associated thunderstorm hazards are not present. Even tornadic storms have been found that produced very little lightning. The lack of electrical activity, as with the absence of a precipitation display on radar, does not necessarily translate into a smooth ride.

Often the question arises: Which is better, radar or lightning displays? Lightning strokes become numerous during the development stage of a thunderstorm with the rapid movement of air and water within the cell. However, as soon as the storm is about to dissipate, its worst weather hazards occur (downdrafts, macrobursts, microbursts, and tornadoes). During this period there is a dramatic decrease in the number of lightning strikes. Just because a storm suddenly stops its lightning strokes doesn't mean it's safe.

Many authorities agree that a combination of radar and lightning detection systems is the best thunderstorm avoidance arrangement. It cannot be overemphasized that these are avoidance, not penetration, devices. Thunderstorms imply severe or greater turbulence and neither radar nor lightning detection systems, at the present, directly detect turbulence.

National Convective Weather Forecast

A new graphic storm-information forecast product for commercial and private pilots is now available through the National Weather Service and the Federal Aviation Administration. Produced by the Aviation Weather Center, the new forecast product provides pilots with a plotted map depicting the current location of convective hazards and where they will be within the next hour.

The National Convective Weather Forecast (NCWF) combines weather service radar mosaics and cloud-to-ground lightning data into a six-color depiction of hazardous weather. An example is in Fig. 2-14. The advanced storm information is designed to make it easier for commercial and private pilots to avoid convective weather hazards in the contiguous United States. NCWF will be used to

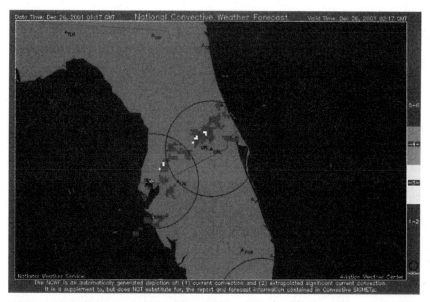

Fig. 2-14. The National Convective Weather Forecast (NCWF) combines weather service radar mosaics and cloud-to-ground lightning data in a six-color, hazardous-weather depiction.

complement on-board radar systems that detect small-scale storms that are less hazardous to aviation. Available on the Internet at http://cdm.aviationweather.noaa.gov/ncwf, and on weather service information networks, the NCWF is updated every 5 minutes.

According to Aviation Weather Center acting director Jack May, "The Aviation Weather Center has been running this forecast as an experimental product for the past 16 months. We anticipate the NCWF will be a great value to pilots in planning and executing their flight routes by showing the quickest and easiest ways to avoid turbulent weather. The National Convective Weather Forecast is now a full-fledged and reliable aviation weather forecast product. Pilots, federal aviation weather briefers, air traffic control specialists, and airline dispatchers who routinely make operational decisions associated with thunderstorm hazards will use the NCWF."

According to May, the NCWF works very well with long-lived mature storm systems, which it was designed to detect. It will also filter out brief, small-scale storms that are not a hazard to aviation or are not likely to persist for an hour. Onboard radar equipment and weather service radar images help pilots and controllers detect and avoid those small-scale storms.

FAA's integrated product team leader for weather and flight service systems, Don Stadtler, said, "As a private pilot, I greatly appreciate the value the NCWF adds to my decision-making process. Its timeliness and ability to help narrow down airspace that I should try to avoid because of potentially hazardous thunderstorms and turbulence are extremely valuable to me and other pilots."

The NCWF forecast product does well with long-lived mature systems. However, the initiation, growth, and dissipation of these systems, as well as shorter-lived isolated storms, are not well forecast. Work on automated methods to forecast the growth and dissipation of storms is ongoing.

The NCWF product is available on a national, ARTCC-specific, and certain airport-specific scales. Figure 2-14 is an airport-specific scale for Orlando, Florida. The time is 0117Z and the image depicts the forecast for 0217Z, one hour (the suffix Z indicates Coordinated Universal Time). The scale on the right depicts standard intensity levels of 1 and 2 (green), 3 (yellow), 4 (orange), and 5 and 6 (red).

Both May and Stadtler praise the product. Certainly it will be of some value to pilots and dispatchers with access to the Internet who are planning an immediate departure to have a 1-hour forecast of convective activity. However, the product is not available on DUATs or to Flight Service Stations or other air traffic controllers, at least at present. Thus the product is of no value to the users of these services. It is certainly not a substitute for a standard weather briefing. In fact, the display specifically states: "The NCFW…does NOT substitute for, the report and forecast information contained in Convective SIGMETs."

CHAPTER 3

Radar Weather Products

Pilots have access to a variety of radar-derived weather products. Of these, pilots are most familiar with the National Radar Summary Chart and Convective SIGMETs. In addition to these products Radar Weather Reports (RAREPs) or SDs (storm detections), the National Reflectivity Mosaic, Radar Coded Message Images, Alert Weather Watches and Bulletins, and some Center Weather Advisories relay information on hazardous convective activity.

SIGMETs were issued for convective activity until a DC-9 crashed in a severe thunderstorm near Atlanta in 1977. From this accident the Convective SIGMET (WST) evolved. The National Aviation Weather Advisory Unit (NAWAU) in Kansas City had WST responsibility. Specifically assigned meteorologists issue these advisories. In October 1995, NAWAU became the Aviation Weather Center (AWC). On Tuesday March 23, 1999, after 110 years in downtown Kansas City, AWC moved to its new quarters near the Kansas City International Airport.

Center Weather Service Units (CWSUs) were established at Air Route Air Traffic Control Centers (ARTCC) in 1980. The purpose of the CWSU is to assist controllers and flow control personnel and to alert pilots of hazardous weather through a Center Weather Advisory (CWA). As is often the case with government bureaucracy, "the cart came before the horse." There were no instructions for ATC personnel when CWAs first appeared. Distribution methods ranged all the way from immediate broadcast to the trash can. The FAA took months to decide that the CWA had the weight of a SIGMET and apply identical distribution and broadcast procedures.

Alert Weather Watches (AWWs) and Severe Weather Watch Bulletins (WWs) are also produced. The Severe Local Storms (SELS) office in Kansas City issued AWWs and WWs for severe thunderstorms and tornadoes until 1997. At that time SELS was relocated to the Storm Prediction Center (SPC) in Norman, Oklahoma.

With the number of advisories, it would seem impossible to fly into an area of hazardous weather without warning. But this is not necessarily the case. An advisory cannot be issued for each individual thunderstorm. Severe weather can develop before an advisory is written and distributed. The absence of an advisory is no guarantee that hazardous weather does not exist or will not develop.

> **CASE STUDY**
>
> When we were returning from Oshkosh, thunderstorms were forecast in the Colorado area. Sure enough, as we approached Pueblo there were several cells. A check with Flight Watch indicated Pueblo was in the clear and Colorado Springs would make a suitable alternate, should Pueblo's weather deteriorate. By keeping visual contact with the storms, we circumnavigated to the north around the heavy rain, lightning, and buildups, and made an uneventful landing. There really is no reason for getting caught in a cell.

The lack of an advisory does not guarantee the absence of hazardous weather. An unfortunate pilot learned this lesson the hard way.

> **CASE STUDY**
>
> The synopsis described a moist unstable air mass. Thunderstorms were not forecast for the time of flight, but were expected to develop; thunderstorms, however, were already being reported along the route. The pilot, without storm detection equipment, encountered extreme turbulence on inadvertently entering a cell.

The pilot, with three passengers, filed an IFR flight plan based on the fact that there were no advisories. About a half hour into the flight, according to the pilot's statements to the FAA and National Transportation Safety Broad (NTSB) from the NTSB report, "…we noticed a heavy layer of clouds at and below our altitude and some 20 miles ahead…. The layer in front of us seemed to be light cumulus with a heavier layer behind it (not ominous looking)." After the encounter the pilot could not understand why he was "never given a precaution or advisory regarding that system!" He went on to say that the accident "would not have happened if (the) pilot had been aware of weather conditions…." There were no advisories in effect because, at the time of the briefing, none were warranted. The pilot had the clues—moist unstable air; thunderstorms already reported—but put complete trust in a forecast that included no precautions or advisories.

The preceding examples illustrate two good decisions. One resulted in a routine flight, the other an almost fatal flight. The intent here is not to criticize the decisions but to show the process on the basis of available information, a knowledge of weather products, and limitations, that led to the decisions.

Automated Radar Weather Reports (RAREPs)

The National Weather Service routinely takes radar observations. These observations are coded and transmitted over the FAA's weather distribution system and are available through DUATs, and the Internet from the Aviation Weather Center's Internet site at www.aviation-weather.noaa.gov. NWS NEXRAD radar site locations are contained in Appendix B, "NWS Weather Radar Chart/Locations." Aerial coverage of these radar sites is graphically depicted in the *Aeronautical Information Manual* (AIM) and AC 00-45 *Aviation Weather Services*. The radar report or SD (storm detection) is now automatically generated by NEXRAD radars, which explains some apparent inconsistencies. For example, in the west, ground clutter was sometimes reported as weather echoes.

> **CASE STUDY**
> In the early days of automated SDs, ground clutter was a
> significant problem in the western United States. In fact, these
> products were all but useless. About the only interpretation an
> FSS briefer or pilot could make from the report was: "Intense to
> extreme terrain existed along the route of flight." Fortunately,
> most of these problems have been eliminated, rendering this
> product a useful tool in determining the amount and intensity of
> precipitation within the radar's area of coverage.

SDs contain the following information:

- Location of the radar

- Time of observation

- Configuration of echoes

- Coverage of echoes

- Type of precipitation

- Intensity of precipitation

- Location of echoes

- Movement of echoes

- Height of echoes

The following illustrates a coded SD report. All reports begin
with the radar site. This is a report for Davenport, Iowa (DVN). Note
that location identifiers (LOCIDs) do not necessarily correspond
to the LOCID used with airports and navigational aids (NAVAID).
This is the 1535Z observation. (This report was observed on
May 17, 1999.)

```
DVN 1535 LN 9TRWXX 95/72 119/119 10W C2237
AREA 1TRW++4R- 288/127 91/114 155W C2239 MT 450 AT 108/99
AUTO
^IJ10112 JI32222222 KI322213332 LI324212344 MH12220001456
NH4302 NQ266 OH45 OR14 PI1 RN1
```

Echo Configuration and Coverage

Echo configuration falls into one of three categories: cell, area, or line. A single isolated area of precipitation, clearly distinguishable from surrounding echoes, constitutes a cell. The following illustrates how cells are indicated on an SD.

FWS 1035 CELL TRWXX 323/95 D30 C2730 MT 550 HOOK 321/91...

This Fort Worth, Texas (FWS) 1035Z observation reports a cell with a 30-nm diameter (D30) exhibiting a hook echo. Cell diameter refers to precipitation, not necessarily the diameter of the cloud, which could be considerably larger. Recall that a hook echo is the signature of a mesocyclone, often associated with severe thunderstorms that produce strong wind gusts, hail, and tornadoes. An area consists of a group of echoes of similar type that appear to be associated. A line (LN) defines an area of precipitation more or less in a line—straight, curved, or irregular—at least 30 miles long, 4 times as long as it is wide, with at least 25 percent coverage within the line. Echo coverage is reported in tenths. In the DVN example, the line has 9/10 (LN "9"TRWXX) coverage and the area has 1/10 (AREA "1"TRW++4R-) coverage of thunderstorms with very heavy rain and 4/10 (AREA 1TRW++"4"R-) coverage of light rain.

Precipitation Type and Intensity

SDs, at least at this writing, use the old, pre-1996, aviation weather symbols. These are shown in Table 3-1. In the DVN example, the line and the area contain thunderstorms and rain showers (TRW). Following precipitation type, one of the six standard levels describe intensity. Intensity level definitions and descriptions are contained in Table 1-2. The line is extreme (TRW"XX") with level 6 intensity echoes. The area consists of isolated (level 4), very heavy (TRW "++") intensity echoes, and scattered (level 1) light-intensity (R"-") precipitation.

One use of the SD is to determine atmospheric stability. This can be inferred from the character of the precipitation. Rain (R) and snow (S) indicate a stable air mass. Rain showers (RW), snow showers (SW), and thunderstorms (T) indicate an unstable air mass. DUAT briefings

TABLE 3-1. RAREP/Radar Summary Chart Plotted Data

SYMBOL	MEANING	SYMBOL	MEANING
PPINE	No echoes	AUTO	Automated Report
PPINA	Not available	T	Thunderstorm
PPIOM	Out for maintenance	R	Rain
NE	No echoes	RW	Rain showers
NA	Not available	S	Snow
OM	Out for maintenance	SW	Snow showers
LM	Little movement		

include the textual SD reports for the pilot's proposed route. A quick scan of the character of the precipitation reveals the stability of the atmosphere. However, since this is an observation and not a forecast, it must be used with other reports and forecasts, specifically the Area Forecast (FA) or Transcribed Weather Broadcast (TWEB) synopses.

Echo Location and Movement

The SD defines the location of precipitation by points, azimuth (true), and distance (nm) from the reporting station. The line in the DVN report extends from a point 95° at 72 nm (95/72) to a point 119° at 119 nm (119/119), 10 nm wide (10W). The points 288° at 127 nm and 91° at 114 nm, 155 nm wide (288/127 91/114 155W), encompass the area. Cell movement indicates short-term motion of cells within the line or area—not necessarily the movement of the line or area itself, which can be considerably different. Cell movement within the line is from 220° at 37 knots (C2237) and area cell movement is from 220° at 39 knots (C2239).

Line or area movement indicates the long-term progress of the system; cell movement indicates short-term motion of cells within the line or area. Before automated reports became available, the weather radar specialist determined line or area movement and indicated this on the SD, which in turn was transferred to the Radar Summary Chart. Unfortunately, this valuable parameter has been lost to automation.

Echo Height

Maximum heights are reported in relation to azimuth and distance from the reporting station, with approximate elevation in thousands of feet above mean sea level (MSL). In the DVN example, maximum tops at 45,000 ft MSL are located 108° from the station at 99 nm (MT 450 AT 108/99). Tops within a stable air mass are usually uniform and may be indicated by the letter U (MT U120, uniform tops to 12,000 ft MSL). It's important to remember that these are precipitation tops, not cloud tops. Precipitation tops will be close to cloud tops within building thunderstorms. However, precipitation in dissipating cells will normally be several thousand feet below cloud tops. The contraction MTS is used to indicate that satellite data as well as radar information was used to measure precipitation tops.

The contraction AUTO is used to indicate that the report is automated, using data from a WSR-88D weather radar site.

RAREP Digital Data

RAREP digital data appears at the bottom of the report. A grid centers on the reporting station. Each block, 22 nm on a side, is assigned the maximum intensity level observed. The maximum intensity level may cover only a small portion of the block. When 20 percent of a block contains light intensity, that level is assigned. Therefore, from digital data alone, all that can be concluded from intensity level 1 is that at least 20 percent of that grid contains light precipitation.

Letters represent coordinates; numbers indicate the maximum intensity level for that coordinate and succeeding coordinates to the right. Refer to Fig. 3-1. In the Davenport example, the first set of coordinates is IJ10112. Locate IJ in Fig. 3-1. This box contains level 1 echoes. The box to the right, IK, does not contain any significant precipitation and is therefore coded with a zero. The next two adjacent boxes to the right are coded as level 1, and IN contains level 2 echoes. These intensities are a prime ingredient in the Radar Summary Chart and Convective SIGMETs.

Figure 3-1 illustrates the Davenport storm detection. The dark-gray area represents the line; the light-gray area depicts the area. Not surprisingly, maximum tops are located within the line. Cell movement

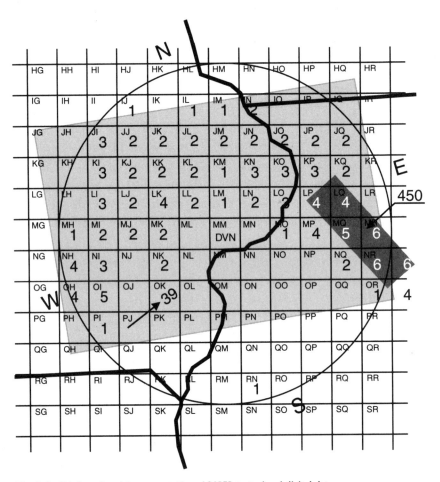

Fig. 3-1. This is a pictorial representation of RAREP textual and digital data.

with the area is from the southwest at 39 knots, as indicated by the arrow. We see that only isolated thunderstorm activity exists within the area. It appears that most significant cells can be circumnavigated with storm avoidance equipment. We should be able to confirm this with actual real-time radar. The line, on the other hand, is not penetrable. In fact, from our knowledge of convective activity, severe weather can extend well beyond the area depicted as the line. Also note that the line may very well extend to the southeast, beyond the coverage of the DVN radar site. We will need a composite radar product or adjacent SDs to determine the actual extent of the line.

Below is another example of a RAREP. This observation came from the Sacramento, California (DAX), NWS radar site. Sacramento radar was not showing any significant weather prior to the following observation.

    DAX 0135 CELL TRW++ 349/80 D9 C2920 MT 250
    AREA 1R- 10/115 46/30 80W C2920 MT 130 AT 33/79
    IN41 KM101=

Look what popped up: an intensity level 4 (very heavy) cell. Thunderstorms were not forecast. This single cell with a diameter of 9 miles and approximate precipitation tops to 25,000 ft "didn't read the forecast." The cell should be easily circumnavigable, and would present a hazard only if it occurred in the vicinity of the departure or destination airport. A pilot should not fly into, close to, or under this cell.

    DAX 0235 CELL RW 346/81 D6 C2820 MT 170
    AREA 1R- 6/95/87/15 80W C2820 MT 120 AT 22/89
    IN21 LO1=

One hour later, the cell has deteriorated to moderate rain showers, with precipitation tops down to 17,000 ft MSL. Convective activity can develop and dissipate, often unforecast, at an alarming rate.

Radar Composite Products

Radar composite products consist of the Radar Summary Chart, the National Reflectivity Mosaic, and the Radar Coded Message Image. All are automated products based on RAREP digital data and graphically display precipitation. The Radar Summary Chart and Radar Coded Message Image are annotated with echo movement and tops. The Radar Summary Chart depicts active severe thunderstorm and tornado watches; the Radar Coded Message Image uses data from Convective SIGMETs and depicts active Convective SIGMETs. The Radar Coded Message Image has a loop capability. The Radar Summary Chart is available through Flight Service Stations and commercial vendors; the National Reflectivity Mosaic and Radar Coded Message Image—along with the Radar Summary Chart—are available on the Internet. These products are all old by the time they are compiled and distributed. Therefore, their use is limited to general flight planning and cannot be substituted for real-time weather radar information.

Most commercial vendors of weather products produce a composite radar summary. An example is contained in Fig. 2-4. Recall that Fig. 2-4 was an FSS National Radar Summary Chart. [For the last 10 years, the FAA has been contracting with various vendors for graphical weather products. Originally, the Operation and Supportable Implementation System (OASIS)—which is supposed to replace existing FSS computer and graphics systems—had its own graphic capability. It remains to be seen just what products will be available with OASIS. By the way, OASIS was supposed to be operational over 5 years ago. This just proves the FAA will not put off till tomorrow what it can put off to the day after tomorrow!] In any case the limitations on the uses of commercial and government radar summary products are the same. However, there are typically some differences in the way data are displayed.

Radar Summary Chart

The Radar Summary Chart graphically displays a computer-generated summation of RAREP digital data. The date and time of the observation—time is important because the transmission system might make the report several hours old—appear on the chart. Figure 3-2 illustrates a May 17, 1999, summary, based on 1435Z data. [This is the same date and approximate time as the Davenport, Iowa (DVN) RAREP previously discussed.] Like the RAREP, the chart contains information on precipitation type, intensity, configuration, coverage, tops and bases, and movement.

Echo movement and tops are depicted by symbology similar to that in the RAREP. An arrow with the speed printed at the arrowhead represents echo or cell movement. For example, echoes within the line in Illinois (Fig. 3-2) are moving from the southwest at 48 knots. Maximum precipitation tops are 46,000 ft. When bases can be determined, the height MSL will appear below the horizontal line.

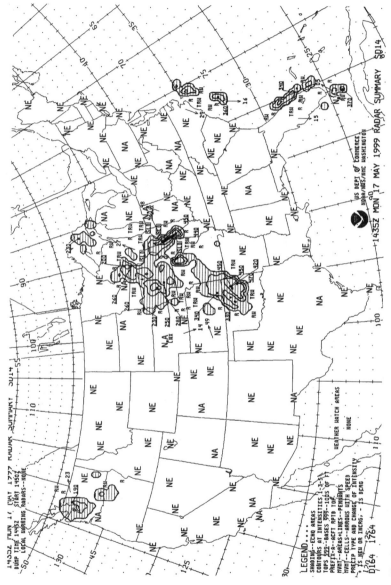

Fig. 3-2. The Radar Summary Chart is a computer analysis of RAREP digital data. It is always old and should be used for preplanning purpose only; always update the radar summary with current observations.

65

Echo configuration is graphically depicted. Echoes reported as a line are drawn and labeled solid (SLD) when at least 8/10 coverage exists; an example is the line in Illinois. From the DVN RAREP data we can determine that this line has 9/10 coverage. The computer plots lines of equal value to indicate echo coverage and intensity. However, unlike the RAREP, the chart displays only three intensity contours. The first contour includes intensity levels 1 and 2 (echoes in central Nebraska and eastern Washington state), the second contour depicts levels 3 and 4 (echoes in western Washington state) and the third contour levels 5 and 6 (echoes in Iowa, northeast Missouri, and around the line in Illinois).

AC 00-45 *Aviation Weather Services* states: "When determining intensity levels from the Radar Summary Chart, it is recommended that the maximum possible intensity be used." Precipitation type uses the same symbology as the RAREP.

When a tornado watch (WT) or severe thunderstorm watch (WS) is in effect, it is listed on the chart to the right of the legend. On this chart under WEATHER WATCH AREAS, NONE is reported; there are no watches in effect at the time of this observation. The location of an active watch will also be depicted as a dashed-line box on the chart.

As with the RAREP, it's important to note which radars are reporting signals that are not echoes (NE) and which radar sites are not available (NA). For example, in Fig. 3-2 there are two radar sites that are not available in southern California. However, the surrounding sites are all reporting no echoes. Using the satellite image and other products, we could confirm that there is indeed no precipitation occurring in this area. Refer to the area of precipitation in the Midwest. There are three radar sites that are not available in Nebraska, eastern South Dakota, and southern Minnesota. We must view these areas with suspicion. They are adjacent to an area of severe weather. It is quite possible that there may be significant precipitation within these areas. Again, we could verify this with other weather products.

Recall that the observations for the Davenport SD depicted in Fig. 3-1 and the Radar Summary Chart in Fig. 3-2 were taken on the same day within an hour of each other. Note that the two correlate very well. From the Radar Summary Chart we see that the line does in fact extend beyond the DVN SD coverage area. From the limited information available, it appears that circumnavigating the line to the south would be the best course of action. From the chart we see that significant precipitation extends well to the west of Davenport, throughout the state of Iowa, and into eastern Nebraska.

National Reflectivity Mosaic

The National Reflectivity Mosaic is available through the Aviation Weather Center's Internet site. The mosaic is created each half hour, from observations taken at around 15 and 45 minutes past the hour. The mosaic is composed of the highest observed reflectivity category within map grid boxes approximately 5 nm on a side. It is designed to give an overall picture of the position, movement, and evolution of precipitation on a large scale. The National Reflectivity Mosaic is illustrated in Fig. 3-3. Note that Fig. 3-3 is the February 16, 2001, 2145 UTC observation.

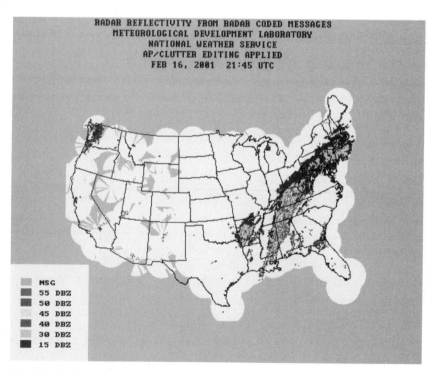

Fig. 3-3. The National Reflectivity Mosaic indicates the approximate intensity of precipitation within the lowest 10,000 feet of the atmosphere.

The reflectivity categories indicate the approximate intensity of precipitation within the lowest 10,000 ft of the atmosphere. Gray areas depict areas more than about 120 nm from the nearest reporting radar, or which are shadowed by terrain features from any reporting radar. This can be seen in the western United States in Fig. 3-3 (the light-gray shaded areas).

Weather radars generally detect many nonprecipitation features. However, automated quality control generally removes ground clutter echoes near the radar and anomalous propagation (AP).

Because each grid box in the mosaic contains the highest reflectivity observed within that box, precipitation areas that appear continuous in the graphic may actually contain many "holes" in the precipitation field—similar to the Radar Summary Chart. Light snow often returns too little energy to appear in the mosaic. Moderate and heavy snow generally do appear.

Despite the automated quality control, nonprecipitation features sometimes appear. These are most often due to migrating birds, and appear during nighttime in the spring and late summer through autumn. Occasionally real precipitation features are removed.

Figure 3-4 is a radar observation for Jackson, Mississippi (JAN), taken on February 16, 2001, at 2149Z. From this observation, we can see an intense line of thunderstorms extending from the vicinity of McComb, Mississippi, through Meridian, Mississippi. Note the very steep precipitation gradient on the southeast side of the line. This would be an extremely high no-go indicator for a flight from Mobile, Alabama to Natchez, Mississippi. The question becomes: Could the pilot circumnavigate the line either to the north or south? Reviewing the National Reflectivity Mosaic (Fig. 3-3) we can see that the convective activity does in fact diminish in Louisiana, between New Orleans and Baton Rouge. However, to the northeast, beyond the radar range of Jackson, the activity continues as an intense line all the way into the Ohio River Valley.

We cannot make a flight decision based solely on the information in the example. However, it does illustrate the ways these products can be used to assist in an informed flight decision.

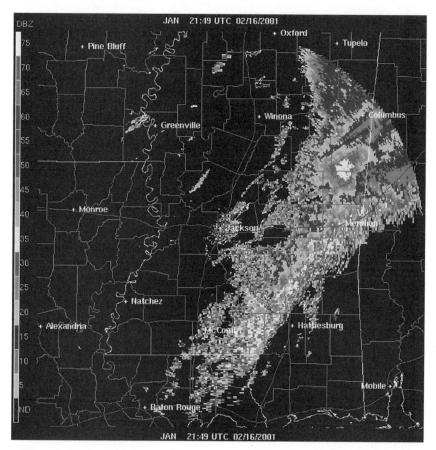

Fig. 3-4. This observation, an intense line of thunderstorms extending from the vicinity of McComb, Mississippi, through Meridian, Mississippi, shows an extremely high no-go indicator for a flight from Mobile, Alabama, to Natchez, Mississippi.

Radar Coded Message Image

Radar Coded Message Images consist of the NEXRAD radar reflectivity graphical overlays of cloud tops and valid Convective SIGMETs. The display is intended to show a national summary of convective activity related to aviation operations.

The NEXRAD Radar Coded Message (RCM) product originates as an automated text message generated at each NWS radar. The RCM reflectivity data has a resolution of about 6 nm and is updated every 30 minutes and posted at H+35 and H+55 (35 minutes and 55 minutes past each hour). The RCM was intended as a replacement for the

Manually Digitized Radar (MDR) product which has been used for years as the input to the NWS Radar Summary Chart. The raw RCM data has considerable false echoes, so it is edited hourly by the Aviation Weather Center (AWC) before being formatted as an automated RAREP product. This automated RAREP is now used as the input for the NWS Radar Summary Chart. The editing of the RCM at the AWC is an automated process that checks the RCM for meteorological validity when compared to current satellite images, synoptic conditions, neighboring radar sites, and lightning data.

The displayed image is the edited reflectivity. The RCM includes the maximum top for each radar's area of coverage. Other tops shown on the display are derived from satellite images at the centers of convective activity. Movements shown are averages generated by the NEXRAD processor.

Figure 3-5 illustrates the Radar Coded Message image. Symbology is similar to that in the Radar Summary Chart. Note the area off the Carolina coast. These echoes have maximum tops of 54,000 ft (*540*). The echoes are moving generally from the southwest at 17 to 18 knots. Note the area in west central Mississippi. According to the chart, echo tops are at 56,000 ft, but echo intensity is only level 1. Certainly this is not convective activity. Most likely it's a band of high clouds with sufficient reflectivity to be evaluated as level 1 by the radar.

The Radar Coded Message Image has a loop capability. Be careful! The loop typically contains the last 6 hours of the product. Because the time frame is so long into the past, its uses are limited. (This is also typical of radar loop images seen on TV.)

Using RAREPs and Radar Composite Products

Pilots can expect to find holes in what the RAREP or Radar Summary Chart portray as an area of solid echoes. This apparent inconsistency is due to several factors. Targets farther from the antenna might be smaller than depicted because of range and beam resolution. NWS weather radars, at a range of 200 miles, cannot distinguish between individual echoes less than 7 miles apart. A safe flight between severe thunderstorms requires 40 miles, so this provides adequate resolution

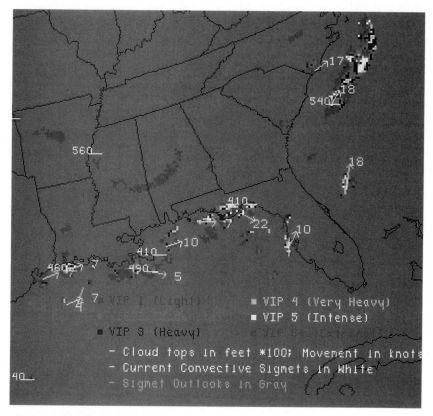

Fig. 3-5. The NEXRAD Radar Coded Message (RCM) product begins as an automated text message with a resolution of about 6 nm, updated every 30 minutes.

to detect a safe corridor. Recall that as little as 20 percent coverage of intensity 1 requires the entire grid to be encoded, so holes also occur with isolated and scattered precipitation. On the Radar Summary Chart, large areas might be enclosed by relatively isolated echoes.

For example, in Washington state west of the Cascades (Fig. 3-2) the chart depicts a large area of light to moderate rain and rain showers, precipitation tops to 19,000 ft, moving from the west-southwest at 23 knots. We would suspect this area to contain only scattered precipitation, except along the west slopes of the Cascades where the orographic effect has resulted in heavy to very heavy precipitation, although the chances of very heavy precipitation are low. How far south does this activity extend? Be careful. The chart shows that the Portland, Oregon, radar is not available (NA).

Assumptions can never be made with RAREPs or the Radar Summary Chart. Can a pilot fly from Oklahoma City and Kansas City while avoiding severe weather by at least 20 miles? (Fig. 3-2) Not without weather avoidance equipment, contact with a facility with real-time weather radar, or visual contact with the convective activity. Why? RAREPs and the Radar Summary Chart are observations, not forecasts. They report what occurred in the past, and convective activity, as we've seen, can develop and move at an astonishing rate. Time of observation is an important consideration. RAREPs might be as much as 2 hours old and the Radar Summary Chart from 2 to 4 hours old. On the other hand, reports from air traffic controllers with access to NEXRAD observations are usually in real time.

Even though the NEXRAD radar network is complete, there are still a few gaps in the west, coverage is designed only for precipitation above 10,000 ft, and ground clutter is still a problem. Although coverage should be adequate for most severe weather, there is no guarantee of complete coverage. Additionally, pilots must be careful to observe the notation NA, which means radar data is not available. This is especially true when you are using products from commercial vendors. What appears to be a hole in convective activity might only be missing data. When using such services, be sure you know how the vendor displays missing data.

Refer to Fig. 3-2. A pilot can never assume there are holes in the solid line in Illinois. The chart can only be interpreted as 8/10 or more coverage within the line. In fact, as we've seen, the Davenport RAREP for about the same time frame showed 9/10 coverage! By reviewing RAREP digital data, a pilot can quickly assess echo intensity and coverage. For example, review the digital data for the Davenport RAREP (Fig. 3-1). Most of the activity is intensity level 1 or 2. But again, without storm detection equipment, access to real-time radar information, or visual contact with convective activity penetrating the area should not even be considered.

The Radar Summary Chart provides general areas and movement of precipitation for planning purposes only, and must be updated by hourly RAREPs or real-time weather radar. Chart notations, such as OM (out for maintenance) or NA (not available) must be considered.

The chart must always be used in conjunction with other charts, reports, and forecasts—the "complete picture." Once airborne, inflight observations—visual or electronic—and real-time weather radar information from FSS or Flight Watch must be used. As stated in AC 00-45E *Aviation Weather Services:* "Once airborne...pilots must depend on contact with Flight Watch, which has the capability to display current radar images, airborne radar, or visual sighting to evade individual storms."

Precipitation should not of itself be cause to cancel a flight. For example, the radar returns in Washington and Oregon in Fig. 3-2 indicate mostly light to moderate activity. Chances are the Cascades and high mountains will be obscured. Again, without further information a flight decision cannot be made with the information available. A flight decision must be made with a knowledge, and consideration, of the following:

1. What is the coverage and intensity of precipitation?

2. What is the weather expected to do (improve/deteriorate)?

3. What is the pilot's experience level, the capability of the aircraft, and the time of day?

It can be difficult to see clouds at night; although they are visible in lightning flashes, distances can be difficult to judge. How familiar are we with the terrain and weather patterns over the intended route? Is an alternate available if the planned flight cannot be completed? Are we mentally prepared to divert, should it become necessary? What about our physical condition? Tired and anxious to get home is a potentially fatal combination.

Convective SIGMETs (WST)

Convective SIGMETs provide detailed, specific forecasts for thunderstorm-related phenomena. Since thunderstorms are accompanied by severe or greater turbulence, severe icing, and low-level wind shear (LLWS), these conditions are not specifically addressed in the advisory or during a weather briefing. The Aviation Weather Center's WST unit makes extensive use of radar data to analyze thunderstorm systems. Meteorologists compare radar data with satellite imagery, lightning information, and other conventional sources to determine the need for a Convective SIGMET. WSTs are issued when the following phenomena occur or are expected to develop and continue for more than 30 minutes within the valid period:

- Severe thunderstorms

- Embedded thunderstorms

- A line of thunderstorms

- An area of active thunderstorms affecting at least 3000 square miles

WSTs for severe thunderstorms may include specific information on tornadoes, large hail, and wind gusts of 50 knots or greater. An embedded thunderstorm occurs within an obscuration, such as haze, stratiform clouds, or precipitation from stratiform clouds. Embedded thunderstorms alert pilots that avoidance by visual means would be difficult or impossible. To be advertised in a Convective SIGMET:

- A line of thunderstorms must be at least 60 miles long, with thunderstorms affecting at least 40 percent of its length.

- Active thunderstorms must have an intensity level of 4 or greater, or affect at least 40 percent of an area.

The 48 contiguous states are divided into three areas for Convective SIGMET issuance: West (MKCW WST, west of 107°W longitude), Central (MKCC WST, between 107°W and 87°W longitude), and East (MKCE WST, east of 87°W longitude). These areas are shown on the Weather Advisory Plotting Chart shown in App. C. Issued on an unscheduled basis as needed, beginning with number 1 at 0000Z, WSTs contain a forecast for up to 2 hours and an outlook for an additional 4 hours.

```
MKCC WST 222055
CONVECTIVE SIGMET 63C
VALID UNTIL 2255Z

MO KS OK
```

```
FROM 40WNW IRK-20WNW BUM-30N TUL
LINE TS 20 NM WIDE MOV FROM 23025KT. TOPS TO FL450.
HAIL TO 1 IN...WIND GUSTS TO 50 KT POSS.
OUTLOOK VALID 222251-230255
FROM TVC-ASP-FWA-SJT-CDS-60WSW PWE-TVC
TS CONTG INVOF CDFNT THAT EXTDS FM CNTRL GRTLKS SWWD
TO DPNG SFC LOW OVER N CNTRL OK. STGST STMS RMN OVER
KS/MO IN AREA OF STGST LOW LVL WARM ADVCTN AND MOST
FVRBL MSTR/INSTBY. STMS EXPD TO INCR IN COVERAGE AND
INTSTY NEXT SVRL HRS. ADDNL STG/PSBLY SEV TS DVLPMT LKLY
INVOF DRYLN THAT EXTDS SWD FM OK LOW.
PDS
```

This central (MKCC) WST was issued on the 22nd day of the month at 2055Z (222055). It is the 63rd central issuance for this ZULU day (CONVECTIVE SIGMET 63C). It affects portions of Missouri, Kansas, and Oklahoma, valid for 2 hours (VALID UNTIL 2255Z).

Specific areas are described by using the VORs on the Weather Advisory Plotting Chart (App. C). In this case they are from 40 miles west-northwest of Kirksville, Missouri, to 20 miles west-northwest of Butler, Missouri, to 30 miles north of Tulsa, Oklahoma. The advisory warns of a line of thunderstorms 20 nautical miles wide moving from 230° at 25 knots, tops to 45,000 ft, hail to 1 inch, and wind gusts to 50 knots possible. The black line in Fig. 3-6 encloses this area; it's nice to have the chart to visualize locations.

The WST outlook was designed primarily for preflight planning and aircraft dispatch. It normally includes a meteorological discussion of factors considered by the forecaster. It is supplemental information not required for weather avoidance, but useful to CWSU and FSS controllers for analysis and background information. Normally the outlook will not be included in broadcasts nor provided during a briefing.

The example outlook is valid for an additional 4 hours, covers an area from Traverse City, Michigan, to Oscoda, Michigan, to Ft. Wayne, Indiana, to San Angelo, Texas, to Childress, Texas, to 60 miles west-

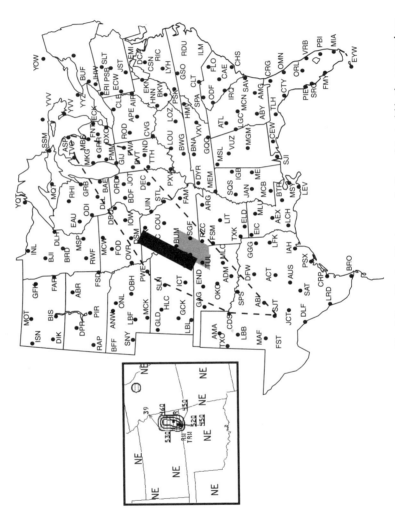

Fig. 3-6. Both the Convective SIGMET and Alert Weather Watch are perfectly consistent within the scope and purpose of each product.

southwest of Pawnee City, Nebraska, back to Traverse City. The broken line in Fig. 3-6 encloses this area.

The outlook translates: Thunderstorms continuing in vicinity of cold front that extends from central Great Lakes southwestward to deepening surface low over north central Oklahoma. Strongest storms remain over Kansas/Missouri in area of strongest low-level warm advection and most favorable moisture/instability. Storms expected to increase in coverage and intensity next several hours. Additionally, strong/possibly severe thunderstorm development likely in vicinity of dryline that extends southward from Oklahoma low.

The outlook interpretation is that thunderstorms will continue because of the cold front. The strongest storms will be produced in the area of strongest low-level warm-air advection (upward-moving air, a lifting mechanism), and most favorable moisture and instability—the three elements required for thunderstorms. Strong, possibly severe thunderstorms are likely to develop along a dryline (a lifting mechanism) that extends from the Oklahoma low-pressure area. Most meteorological terms used in the outlook discussion are contained in the Glossary.

Severe Weather Bulletins

The National Weather Service produces three severe weather bulletins that apply directly or indirectly to aviation operations. The alert weather watch (AWW) warns of severe convective weather. The severe weather watch bulletin (WW) provides details on the AWW. The hurricane bulletin (WH) advertises hazards associated with this phenomenon. Both the AWW and WH are written in plain language, since they are primarily public forecasts.

Alert weather watches (AWWs) alert forecasters, briefers, pilots, and the public to the potential for severe thunderstorms or tornadoes. Subsequent to the AWW, a Severe Weather Watch Bulletin (WW) is issued. The WW contains details on the phenomena described in the AWW. These unscheduled bulletins are primarily a public forecast, whereas the WST is a combination observation and aviation forecast.

Although Storm Prediction Center and Aviation Weather Center meteorologists coordinate their products, criteria and time frames differ. Therefore, aerial coverage might not coincide. The issuance of an AWW might precede or coincide with a WST. AWWs are numbered sequentially beginning each January 1. The following AWW was issued just after the WST in the previous example. (Refer to the gray shaded area in Fig. 3-6.)

```
MKC AWW 222036
WW 166 TORNADO OK KS MO AR 222100Z - 230200Z
AXIS..70 STATUTE MILES EAST AND WEST OF LINE..
10S MKO/MUSKOGEE OK/ - 65NNE JLN/JOPLIN MO/
..AVIATION COORDS.. 60NM E/W /45SSE TUL - 25SE BUM/
HAIL SURFACE AND ALOFT..2 INCHES. WIND GUSTS..70 KNOTS.
MAX TOPS TO 500. MEAN STORM MOTION VECTOR 25025.
```

When the area is described with locations not on the Weather Advisory Plotting Chart, a separate line titled ..AVIATION COORDS.. will be added. The mean storm motion vector or mean wind vector is the direction and magnitude of the mean winds from 5000 ft AGL to the tropopause. It can be used to estimate cell movement. In the example the mean storm motion vector is from 250° at 25 knots.

The WST describes a developing area of interest to aviation. Inset in Fig. 3-6 is a portion of the Radar Summary Chart observed at 2035Z. Thunderstorms have already developed along the southern portion of the line described in the WST. The area is moving from the southwest at almost 40 knots. The WST forecaster expects the line to develop toward the northeast. Additionally, the AWW forecaster expects tornadic activity to develop along the southern portion of the line and move eastward. Expect later WSTs to cover the area toward the northeast and southwest, into the AWW, and along the area described in the WST outlook. All of this is perfectly consistent within the scope and purpose of each product.

Below is the severe weather watch bulletin (WW) corresponding to the alert weather watch (AWW) in the previous example.

MKC WW 222036
URGENT - IMMEDIATE BROADCAST REQUESTED
TORNADO WATCH NUMBER 166
STORM PREDICTION CENTER NORMAN OK
336 PM CDT THU APR 22 1999
THE STORM PREDICTION CENTER HAS ISSUED A
TORNADO WATCH FOR PORTIONS OF
    NORTHEAST OKLAHOMA
    EXTREME SOUTHEAST KANSAS
    SOUTHWEST MISSOURI
    NORTHWEST ARKANSAS

EFFECTIVE THIS THURSDAY AFTERNOON AND EVENING FROM 400PM UNTIL 900PM CDT.

TORNADOES...HAIL TO 2 INCHES IN DIAMETER...THUNDERSTORM WIND GUSTS TO 80 MPH...AND DANGEROUS LIGHTNING ARE POSSIBLE IN THESE AREAS. THE TORNADO WATCH AREA IS ALONG AND 70 STATUTE MILES EAST AND WEST OF A LINE FROM 10 MILES SOUTH OF MUSKOGEE OKLAHOMA TO 65 MILES NORTH NORTHEAST OF JOPLIN MISSOURI.

REMEMBER...A TORNADO WATCH MEANS CONDITIONS ARE FAVORABLE FOR TORNADOES AND SEVERE THUNDERSTORMS IN AND CLOSE TO THE WATCH AREA. PERSONS IN THESE AREAS SHOULD BE ON THE LOOKOUT FOR THREATENING WEATHER CONDITIONS AND LISTEN FOR LATER STATEMENTS AND POSSIBLE WARNINGS.

DISCUSSION...THREAT FOR SEVERE THUNDERSTORMS AND POSSIBLY SUPERCELLS SHOULD INCREASE ALONG COLD FRONT THIS AFTERNOON AS AIR MASS CONTINUES TO DESTABILIZE AND CAP WEAKENS. TORNADO THREAT APPEARS GREATEST IN OK IF ISOLATED CELLS CAN DEVELOP. AVIATION...TORNADOES AND A FEW SEVERE THUNDERSTORMS WITH HAIL SURFACE AND

ALOFT TO 2 INCHES. EXTREME TURBULENCE AND SURFACE WIND GUSTS TO 70 KNOTS. A FEW CUMULONIMBI WITH MAXIMUM TOPS TO 500. MEAN STORM MOTION VECTOR 25025.

...VESCIO

Center Weather Advisories (CWAs)

Center weather advisories, unscheduled inflight advisories, are issued when conditions are expected to significantly affect IFR operations and help pilots avoid hazardous weather. The advisories update or expand the AIRMET Bulletin, SIGMETs, Convective SIGMETs, and the Area Forecast. They may be issued when conditions meet advisory criteria. In such cases, the Center Weather Service Unit will coordinate with Aviation Weather Center forecasters for the issuance of the appropriate advisory. CWAs are also issued when local hazardous conditions develop that do not warrant other advisories. Because they often report localized phenomena, the area might be described using locations other than those on the Area Designators map, or VORs on the Weather Advisory Plotting Chart.

The CWA numbering system was somewhat complex, but has been simplified. CWAs have a three-digit number. The first digit is a phenomenon number, that is, a specific weather event that required the issuance of the CWA. A separate phenomenon number will be assigned each distinct condition (turbulence, icing, thunderstorms, etc.). For example, 101 may forecast a turbulence event; 201 might be issued for icing. The second and third digits indicate the number of times a specific phenomenon event has been updated. For example, 101 (first issuance), 102 (second issuance), and so on.

This Houston Center CWSU advisory is the first issuance for this phenomenon (ZHU CWA 101). It is valid from the 6th day of the month at 1355Z until the 6th day of the month at 1455Z (061355-061455).

ZHU CWA 101 061355-061455

FM A BPT to 40SE LFT LN..S 150 MI INTO GULF...AREA SCT INTST 3-5 TSTMS MOVG N 15 KTS. NMRS TOPS ABV 450.

The Houston Center CWSU has issued this advisory for an area of scattered intensity level 3 to 5 thunderstorms moving north at 15 knots, with numerous tops to above 45,000 ft MSL. The area extends along a line from Beaumont, Texas to 40 miles southeast of Lafayette, Louisiana south 150 miles into the Gulf of Mexico. These locations are not on the Weather Advisory Plotting Chart. The condition has not yet met the criteria for a WST.

CWSUs also issue Meteorological Impact Statements (MISs). Strictly an in-house product, the MIS alerts controllers of weather that might affect the flow of IFR traffic. The MIS describes conditions already contained in other advisories and forecasts. From time to time, overzealous FSS controllers might refer to an MIS or tower controllers might record it on the Automatic Terminal Information Service (ATIS).

Dissemination

Weather advisories are routinely provided during FSS standard briefings and offered during abbreviated briefings. During routine FSS radio contacts, advisories within 150 miles will be offered when they affect the pilot's route. It's important to note SIGMET and WST number to ensure receipt of the latest information.

In the contiguous United States, Hazardous Inflight Weather Advisory Service (HIWAS) has been commissioned. Advisories and urgent PIREPs are broadcast continuously over selected VORs. The availability of HIWAS can be determined from aeronautical charts and the *Airport/Facility Directory*.

When an advisory affects an area within 150 miles of a HIWAS outlet or an ARTCC sector's jurisdiction, an alert is broadcast once on all frequencies—except Flight Watch and emergency. Approach controls and towers also broadcast an alert, but it may be limited to phenomena within 50 miles of the terminal. When the advisory affects operations within the terminal area, an alert message will be placed on the Automatic Terminal Information Service.

In spite of criticism that advisories cover too much area, their issuance has become more conservative. Ironically, some pilots and

> **CASE STUDY**
> Overzealous tower controllers have been known to place SIGMET alerts on their ATIS for conditions hundreds of miles from the airport. At the FSS we would have dozens of pilots calling for the text of these advisories. We looked at it as job security.

FSS controllers now criticize the forecast for not containing enough precautions. Virtually all criticism, however, is due to misconceptions and misunderstanding the product.

The existence of an advisory, or lack thereof, does not relieve the pilot from using good judgment and applying personal limitations. Like all pilots, I have had on occasion to park my turbo Cessna 150 and go by rental car or the commercial airlines. These instances lend credence to the sage aviation axiom: "When you have time to spare, go by air, more time yet, take a jet." When you don't have the equipment or qualifications to handle the weather, don't go! This doesn't mean every time we hear an advisory we cancel; but, we do take a close look at all available information—the "complete picture."

CHAPTER 4

Introduction to Satellite Imagery

Weather observations consist of surface observations—METAR, observations from pilots either on the ground or during flight in the form of pilot weather reports—PIREPs, weather radar, upper air soundings, and satellite images. Our three-dimensional observational system begins with surface observations—the lower layer; next comes PIREPs, weather radar, and upper air observation—the middle layer; finally satellite images provide a look from the top down.

Since April 1960, weather satellites have been orbiting over the Earth. There are two main types of meteorological satellites: polar orbiters and equator orbiters. The polar orbiters range in altitude from about 400 to 600 nm. They circle the Earth in orbits that carry them over the poles. As Earth rotates below them, they scan the entire globe, one strip at a time. Geostationary Operational Environmental Satellites (GOES) orbit directly over the equator at approximately 19,000 nm. They circle the Earth once every 24 hours. From the satellites' view the Earth appears to remain stationary, thus the name *geostationary*.

Pilots usually have access to the Geostationary Operational Environmental Satellites. GOES observations are available through Flight Service Stations (FSSs), including Flight Watch, on the Internet, and are typically those satellite images seen on television. Our discussion will be limited to GOES, although the same principles and limitations apply equally to the polar orbiters. Most of the satellite images used here, and for our discussion, are by courtesy of the National Oceanic and Atmospheric Administration (NOAA). In addition to current images, NOAA has a satellite image archive on the Internet at www.goes.noaa.gov.

GOES East is located at approximately 75° west longitude and GOES West at approximately 135° west longitude. GOES East covers the eastern two-thirds of the United States, the Caribbean, and the Atlantic. GOES West covers the western third of the United States, eastern Pacific, Hawaii, and Alaska.

Satellite interpretation is a science in itself. Therefore, we will limit our discussion to two basic types of image: visible and infrared (IR). Visible, as the name implies, is a "snapshot" of conditions on Earth. A visible image is the result of reflected sunlight. Resolution of visible images ranges between 0.5 and 5 nm. IR is a temperature picture. That is, the satellite senses the temperature of an area with a resolution of approximately 5 nm. (Most of the pictures used in this chapter, and available from the NOAA Web site, have a resolution of approximately 5 nm.) Resolution deteriorates with distance from the satellite, both north and south of the equator and west or east of the satellite's position. This error is known as parallax error. As a result of the limitations of resolution and parallax, some clouds may be displaced several miles from their actual location and some objects will not show accurate brightness values. Cloud elements smaller than the resolution of the satellite will not be detected—another limitation of satellite imagery.

Normally, small or thin clouds, in themselves, do not present a hazard to aviation. However, a problem might arise should a pilot make an incorrect interpretation of a satellite image. It's important to be able to accurately interpret the shading on the image. The problem with small or thin clouds is illustrated in Fig. 4-1.

In an area of small or thin clouds, part of the reflected sunlight sensed by the satellite (represented by the solid lines in Fig. 4-1) is from the tops of the clouds and part from the land or water surface below. The resultant gray shade depicts an average of the two reflectivities. It is darker than a thick cloud and lighter than the normal surface shade as depicted on visible imagery. An exception

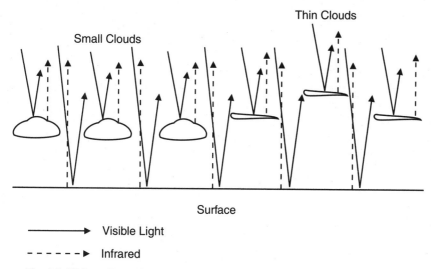

Fig. 4-1 With small or thin clouds, part of the reflected sunlight is from the tops of the clouds, part from the surface below; the infrared sensor averages the temperature of the cloud tops with that of the surface temperature.

would be a surface covered with snow or a thick layer of low clouds. Then there would be no noticeable error in shading, although, under these conditions, small or thin clouds may not be detectable at all.

Small or thin clouds would also be incorrectly depicted on IR imagery. The satellite sensor would average the temperature of the cloud tops with that of the ground temperature. (Infrared radiation is represented by the dashed lines in Fig. 4-1.) As a result, the cloud layer would be portrayed at a level lower than its actual altitude. For example, assume an area is half covered with small clouds. The tops of the clouds are at 10,000 ft, and the temperature at the tops of the clouds is 0°C. The temperature of the surface at sea level is 20°C. Half the radiation coming from that area would be from the tops of the clouds and half from Earth's surface. The satellite would sense an average temperature of 10°C, and the resultant gray shade on the photo would correspond to 10°C.

Most satellite images have a grid depicting geographical and political boundaries. With satellite imagery it is important to establish that the grid location is indeed correct. As we will see this can often be done by comparing the grid with visible geographical features.

> **CASE STUDY**
> Prior to the current generation of weather satellites, the grid was put on the image at the satellite. Well, you can guess what happened; Murphy's law came into play. An error developed in the placement of the grid, and it's a little tough to service these things 19,000 miles above Earth. We were still able to use the images, but a lot of mental gymnastics was required for proper interpretation. The latest generation of satellites places the grid on the image at the receiver site, thus eliminating any such problem in the future.

Visible Imagery

For visible imagery, various types of clouds and terrain reflect different amounts of sunlight. The clouds are white, land masses gray, and water very dark—almost black. At night, when there is no sunlight, there can be no visible imagery. (The United States Air Force, however, has attained some excellent results with moonlight for nighttime visible weather images.) Clouds are excellent reflectors, so they appear very white. The best reflectors are large cumulonimbus clouds. Thin clouds or areas of very small clouds appear darker because less sunlight is reflected. Various types of terrain have intermediate or low reflectivity so land surfaces appear as a shade of gray. Water surfaces, the poorest reflectors, appear almost black. Water will always be very dark unless it is very shallow, muddy, or frozen. Therefore, land-water contrast will typically be very good on visible imagery.

> **CASE STUDY**
> I was briefing a pilot early one morning at the Oakland FSS. The pilot asked for a description of the visible satellite image. In an attempt at a little humor I explained that the "flash bulb" on the satellite had failed. The pilot responded: "Yes, you've been having a lot of trouble with that satellite lately." Oh, well.

Below is a list of the reflectivity of various surfaces by percentage.

- Large thunderstorms, 92 percent

- Fresh new snow, 88 percent

- Thick cirrostratus, 74 percent

- Thick stratocumulus, 68 percent

- White Sands, New Mexico, 60 percent

- Snow 3 to 7 days old, 59 percent

- Thin stratus, 42 percent

- Thin cirrostratus, 32 percent

- Sand, no foliage, 27 percent

- Sand and brushwood, 17 percent

- Coniferous forest, 12 percent

- Water surfaces, 9 percent

We'll refer back to this list in subsequent chapters.

Snow cover is often difficult to identify. Both low clouds and snow reflect about the same amount of sunlight. This is especially true over relatively flat terrain. Differentiating between low clouds and snow can often be done by recognizing terrain features, such as unfrozen rivers and large lakes. Clouds normally obscure terrain features, but snow cover does not. Snow in mountainous areas is usually easier to identify because it often forms a dendritic pattern. Mountain ridges above the tree line are essentially barren, and snow is visible. In the tree-filled valleys, most of the snow is hidden beneath the trees, and a branchy, sawtooth, dendritic pattern identifies areas of snow cover. Dendritic patterns are illustrated in Fig. 4-2. The southern Sierra Nevada Mountains are cloud-free, and the dendritic patterns can easily be seen.

Figure 4-3 is a visible image from GOES East (the GOES 8 satellite) stationed at 75° west longitude. Notice the black background. Neither Earth's atmosphere nor empty space reflect any sunlight, so any background area on visible imagery will always be black. Pilots will seldom need a full disk photo (as seen in Fig. 4-3). Most users want one of the sectors, which will result in higher resolution and more detail. Note the dark area in the center-right portion of Fig. 4-3. This is the sunset line, commonly known as the terminator; it marks the portion of Earth that is in darkness and therefore cannot be seen on

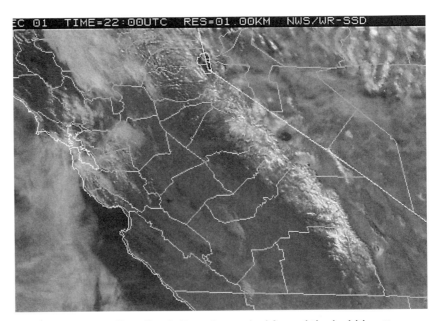

Fig. 4-2 The southern Sierra Nevada Mountains are cloud-free and the dendritic patterns can easily be seen.

visible imagery. Because of the large scale of Fig. 4-3, only major geographical boundaries are portrayed.

On visible imagery, all thick clouds will be essentially the same shade, almost white, regardless of their altitude. Small or thin clouds will appear darker than thick clouds. Differences in shading do not provide any information about cloud height, only cloud thickness.

Infrared Imagery

Everything with a temperature above absolute zero radiates electromagnetic energy. The wavelength of this radiation varies with the temperature. As energy radiates from the surface of Earth and the tops of clouds, the infrared sensor in the satellite measures the energy level at specific wavelengths.

IR images begin by portraying different temperatures as black, shades of gray, and white; black is the warmest, white the coldest. Typically, black represents a temperature of about 33°C and white −65°C, with the gray shades representing decreasing temperature

11 10 2001 1745Z

VISIBLE   NOAA

Fig. 4-3 GOES East (the GOES 8 satellite) is stationed at approximately 75° west longitude over the equator.

toward the white end. In fact, there are 256 distinct shades from black to white. But, basically, warm temperatures are dark, cool temperatures gray, and cold temperatures light.

Figure 4-4 illustrates how an unenhanced IR image is shaded. It is simply a straight-line relationship between temperature and gray shades. Temperature is shown on the bottom along the horizontal axis, and shades between black and white are on the vertical axis. The temperature range covered is from 56.8°C (which might be found in the Sahara Desert in summer) to −109°C (which might be found at the tops of very high clouds—about 60,000 ft). A computer is capable of producing and recognizing 256 distinct shades from black to white.

However, the human eye can distinguish between only about 15 to 20 shades, depending on conditions. Therefore, an unenhanced IR image is of limited operational use. A technique called "enhancement" is used to highlight areas of interest. Enhancements will be specifically covered in subsequent sections.

Land-water contrast is also dependent on temperature. During daylight hours, as the land warms, it takes on a dark-gray shade as its temperature increases. Water areas, however, remain nearly the same temperature and appear as a shade of gray. At night the land cools rapidly and may become cooler than the water. Then the land may appear as a lighter shade than the water. When the temperatures are the same, there is no discernible land-water contrast.

Snow cover is all but impossible to determine from an unenhanced infrared image. Normally, there is not enough temperature contrast between the snow cover and adjacent surface areas to appear as contrasting shades of gray. On days when the sky is clear, barren terrain that is not covered by snow can become significantly warmer than nearby snow-covered areas, at least for a few hours in the afternoon. Under these conditions, the boundaries of snow cover can be identified on the IR image. There are enhancement curves

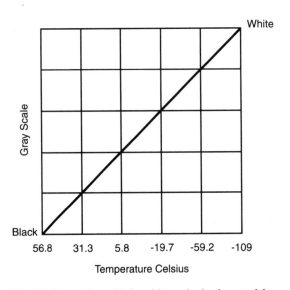

Fig. 4-4 An unenhanced infrared image is simply a straight-line relationship between temperature and gray scale.

developed specifically to show snow cover and snow melt information, but these are seldom used for aviation purposes.

Figure 4-5 is an IR image from GOES West (the GOES 10 satellite) stationed at approximately 135° west longitude. Since Earth's atmosphere and space do not radiate infrared energy at the satellite's sensor frequency, these areas appear white on the image. Notice in Fig. 4-5 that there is virtually no contrast between the land and oceans. On this IR image about the only way to distinguish between the two is by using the superimposed grid.

Keep in mind that all infrared images are nothing more than representations of temperatures. Surface temperatures can vary

Fig. 4-5 GOES West (the GOES 10 satellite) is stationed at approximately 135° west longitude over the equator.

greatly from day to night, so the shading on infrared images will also vary considerably in a 24-hour period.

Enhancement

Computer technology allows for the enhancement of infrared images. This technique allows the operator to highlight areas of interest. In the enhancement process, any shade of gray may be assigned to any temperature when more contrast is needed to highlight a certain temperature range. Colors can also be assigned to specific temperature values within the 256 gray shades. This results in the variety of color satellite images seen on television and available through various Internet sites. However, without knowing the exact "enhancement curve," specific interpretation is difficult. Enhancement curves allow for greater detail of certain phenomena, such as snow and ice, fog and thunderstorms, haze, dust, and volcanic ash. Most, typically, are not available for pilot use in an operational environment. Therefore, our discussion will be limited to those satellite products and phenomena for which pilots normally have operational access.

For an enhancement curve to be effective, two basic criteria are considered. First, a limited number of features are enhanced. Second, the enhancement should contain as much detail as possible without making the display overly cluttered. This technique has the advantage of making deep, vertically structured cloud areas obvious. However, some details may be lost in assigning a range of temperatures to a given gray shade or color. Multiple enhancements provide increased contrast over the total temperature range, but also introduce multiple boundaries in cloud systems, which may mask significant cloud edges. There are always tradeoffs between simple enhancement curves, which sacrifice detail, but can be quickly interpreted for operational use, and complex curves, which maximize the information content, but require more time to interpret.

Figure 4-6 illustrates a slight modification of the unenhanced IR curve described in Fig. 4-4. The line in Fig. 4-4 depicts a simple enhancement curve. The very warm temperatures in segment 1 are all shown as black. Since there are no clouds in this temperature range, there is no need for different shading.

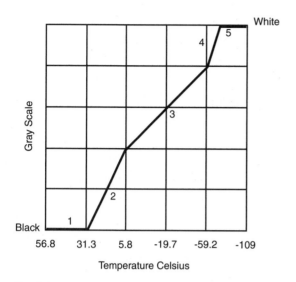

Fig. 4-6 A slight enhancement to the basic IR curve results in a more usable image for depicting cloud cover.

Segment 2 contains the temperatures of most low clouds and water surfaces. This allows the darker shades of gray to be used over a smaller temperature range so that small differences in temperature can be more easily detected.

A similar process is done at the other end of the temperature scale, except that it affects a much smaller range of temperature. All temperatures of −75.2°C or colder are shown as pure white—segment 5. Here, too, since there are so few clouds in this temperature range, there is no need for a difference in shading. The shades from very light gray to white are used in the temperature range of the very high cirrus clouds—segment 4.

Between the two extremes in segment 3, no enhancement is used. Clouds in this temperature range will be shaded the same as on the unenhanced IR image. This curve is a somewhat improved version of the unenhanced curve, and these infrared images are generally used in place of the unenhanced infrared images. A summary of the enhancements is contained in Table 4-1.

Many enhanced IR images have a gray scale directly below the product legend. It is a visual explanation of the enhancement curve. The vertical lines are in 10°C increments of temperature. This

Table 4-1 Enhanced infrared image

SEGMENT	TEMPERATURE, °C	REASON
I	56.8 to 29.3	Little or no meteorological data
2	28.8 to 6.8	Low level/sea surface difference
3	6.3 to −55.2	Middle level—no enhancement
4	−56.2 to −75.2	Upper-level enhancement
5	−75.2 to −110.2	Little or no meteorological data

temperature scale is shaded the same way the image is shaded to enable users to determine at a glance what temperature or range of temperatures is depicted. As previously mentioned, colors can be substituted for any of the 256 shades of gray. This is typically the process used on television weather programs. Unfortunately, we rarely, if ever, have access to the enhancement data used for these products.

There are literally hundreds of enhancement curves. By studying the enhancement curve, if one is available, it can be determined exactly what temperature range or ranges have been enhanced. This will provide an idea of what features are being highlighted and will assist in interpretation. An example of an enhancement curve, along with its interpretation and application, is contained in the final section of this chapter.

Cloud Types

Clouds form through a cooling process. The process initiates and then must sustain condensation or sublimation. (Sublimation is the process where water vapor changes state directly to a solid-frozen state, without first becoming a liquid.) Cooling processes are adiabatic or diabatic.

The adiabatic process cools the air by lifting the parcel through the processes of convection, convergence, or orographic lifting. The diabatic process produces a loss of heat. The loss may occur through terrestrial radiation, resulting in fog or low clouds. Conduction through contact with a cold surface may result in dew, frost, or fog. The process may be associated with the movement of air across a cold surface—advection. Finally, the process may occur through mixing with colder air. If the mixture has a temperature below its dew point, clouds or fog may form.

There are a number of methods of cloud classification. Clouds may be classified according to appearance, how they are formed, or the height of their bases. It was not until 1803 that Luke Howard, an

Englishman, first classified cloud forms. He divided clouds into three main categories using Latin names:

- Cirrus—curly

- Stratus—meaning spread out

- Cumulus—heaped up

    Two prefixes/suffixes may be added:

- Alto—high

- Nimbo—rain

    Today, meteorologists divide clouds into four main groups:

- Low clouds (bases near the surface to about 6500 ft)

- Middle clouds (bases from 6500 ft to 20,000 ft)

- High clouds (based at or above 20,000 ft)

- Clouds with vertical development (based near the surface, tops of cirrus)

Sometimes clouds are classified into one of two general classifications: stratiform and cumuliform. Stratiform describes clouds of extensive horizontal development, associated with a stable air mass. Stratiform clouds consist of small water droplets. The following cloud types are classified as stratiform:

- Stratus

- Stratocumulus

- Nimbostratus

- Altostratus

- Cirrostratus

Cumuliform describes clouds that are characterized by vertical development in the form of rising mounds, domes, or towers, associated with an unstable air mass. Because of upward-moving currents, cumuliform clouds can support large water droplets. In the case of cumulonimbus, updrafts can support hail. The following cloud types are classified as cumuliform:

- Cirrocumulus

- Altocumulus

- Cumulus

- Cumulonimbus

A third generic cloud type, in addition to the two previously mentioned general classifications, is often used. It describes the entire group of high clouds: cirroform. Cirroform is often used during pilot weather briefings and in aviation weather products to translate one or all of the high cloud types: cirrus, cirrostratus, or cirrocumulus.

Texture is one means of identifying cloud types. Texture is caused by shadows. Stratiform clouds appear flat and sheetlike because they are formed in stable air. Stratiform clouds normally show no texture. Cumuliform clouds appear rounded, billowy, and puffy on visible imagery because they are formed in unstable air. Cumuliform clouds have a lumpy texture. Cirroform clouds often have a fibrous texture. Cloud appearance is an excellent indicator of atmospheric stability.

Often in unstable air, the tops of some cumulus clouds are higher than others. Early in the morning and late in the evening when the sun angle is low, the higher tops cast shadows on lower clouds or the surface. These shadows can be seen on visible satellite imagery and result in a texture pattern. At midday, when the sun is directly overhead, no shadows are cast, and there will be no texture. Even when there are no shadows, cumuliform clouds may still be recognizable by their lumpy appearance. Sometimes, however, cumuliform clouds may look very much like stratiform clouds on the visible image.

Stratiform clouds normally do not have texture because of their flat tops. However, when there is more than one cloud layer, the higher layer may cast shadows on a lower layer or the surface.

Shadows do not appear on infrared imagery. Therefore, IR images lack any texture. However, on enhanced IR images contours may appear where the shadows would be on the visible picture. These contours show the edges of the higher clouds very clearly when compared to the shade of the lower clouds.

Low cloud tops, stratus, and fog are characterized by a flat, smooth, white appearance, and a lack of an organized pattern or texture.

Boundaries are often sharp and defined by topography and may exhibit a dendritic pattern. It may be difficult to distinguish fog from low stratus and snow. On visible imagery they appear bright when thick; on IR images they appear dark to medium gray and may be difficult to distinguish from the surface since there is little temperature difference between the surface and cloud tops. Most types of fog show up well on visible satellite imagery. Often, the satellite is better at determining the extent of fog and stratus than surface observations. Stratus indicates a stable air mass. Precipitation, when it occurs, is usually light, often in the form of drizzle.

Stratocumulus clouds represent a moist layer with some convection. Stratocumulus may form from the spreading out of cumulus, which indicates decreasing convection. Stratocumulus can develop from stratus with winds of moderate to strong intensity. Stratocumulus clouds appear bright on a visible image with some texturing; on an IR image they are typically medium to dark gray.

Stratocumulus clouds often appear in sheets or lines of clouds. Sometimes individual cloud elements are seen. A cloud element is the smallest cloud that can be seen on satellite images. These clouds sometimes form in narrow bands in which individual cloud cells are connected (and are known as cloud lines) or not connected (known as cloud streets). Stratocumulus sometimes exhibits a cellular cloud pattern; that is, it forms in more or less a pattern of cloud cells. It may form a closed cell or open cell pattern. *Closed cell* refers to the fact that cloud cover is solid, with individual convective elements rising through the layer. *Open cell* indicates clear air surrounding each individual convective cell. On visible imagery, cumuliform clouds typically appear rounded.

Nimbostratus clouds are low, usually uniform, and dark gray in color. Nimbostratus usually evolves from altostratus that has thickened and lowered, sometimes with a ragged appearance. This is the ordinary rain cloud that produces light to moderate steady precipitation. Nimbostratus appears very bright on a visible image with no texturing; on an IR image, it is typically light gray to white because of their relatively high tops and cold temperatures.

Middle clouds fall into two general types: altostratus and altocumulus. Altostratus indicates a stable atmosphere at middle levels.

Some altostratus clouds are thin and semitransparent, while others are thick enough to hide the sun or moon. Altostratus often indicates the approach of a warm front. These clouds can produce precipitation in the form of rain or snow, even heavy snow, at times. Altocumulus indicates vertical motion and instability at middle levels. Altocumulus may be thin, mostly semitransparent. Some altocumulus clouds are thick, developed, and may be associated with other cloud forms. This cloud often signals the approach of a cold front. Altostratus appears bright on a visible image, with no texturing; altocumulus also appears bright on a visible image, with texturing. On an IR image both are typically light gray to white because of their high tops and cold temperatures. Middle clouds may appear in cellular (altocumulus) or sheet (altostratus) patterns. Like stratocumulus, altocumulus may exhibit stationary lines (altocumulus standing lenticular pattern) indicating mountain wave activity. On IR imagery these clouds appear lighter than low-level clouds because of their colder tops.

High clouds are the cirrus, cirrostratus, and cirrocumulus families. Cirrus clouds are composed entirely of ice crystals. Usually the air is so cold that they do not present an icing hazard. Cirrus often consists of filaments, commonly known as mares' tails. Other cirrus is associated with cumulonimbus clouds. A thickening cirrus layer may indicate the approach of a front. Cirrostratus describe sheets or layers of cirrus. Sun or moon halos may occur. When cirrostratus appear within a few hours after cirrus in midlatitudes, there is a good probability of an approaching front. Cirrocumulus indicates vertical motion at high levels. Cirrus, of itself, has no significance to low-level flights. Thick cirrus appears bright on a visible image with no texturing, except for cirrocumulus; on an IR image it is typically white. An exception would be thin cirrus, which appears gray on a visible image but white on an IR image because of its extremely high, cold tops.

High clouds (cirroform) form where temperatures are very cold. Thin cirrostratus appear on visible imagery as a medium gray, on IR imagery as light or very light gray. Recall the discussion of thin clouds. Thick cirrostratus appears on visible and IR imagery as almost white. Thick cirrostratus is among the highest in reflectivity, with very cold tops. As we shall see, one of the best techniques to identify cirrus is by comparing both visible and IR images of the same time frame.

Clouds with vertical development are cumulus and cumulonimbus. Some cumulus describes fair weather. Other cumulus contains considerable vertical development, generally towering. This type precedes the development of cumulonimbus and thunderstorms. METAR reports may use TCU (towering cumulus) to describe this cloud. This refers to growing cumulus that resembles a cauliflower, but with tops that have not yet reached the cirrus level. With cumulus clouds, expect showery, often heavy precipitation. Cumulus and cumulonimbus appear bright on a visible image with considerable texturing; on an IR image they are typically white, unless their tops are at a low altitude with relatively warm temperatures.

Cumulonimbus clouds exhibit great vertical development with tops composed, at least in part, of ice crystals. Tops no longer contain the well-defined cauliflower shape of towering cumulus. Cumulonimbus may develop a clearly fibrous (cirroform) top, often anvil shaped. Regardless of vertical development, a cloud is classified as cumulonimbus only when all or part of the top is transformed, or in the process of transformation, into a cirrus mass. Any cumulonimbus cloud should be considered a thunderstorm with all the ominous implications. METAR reports and Terminal Aerodrome Forecasts (TAFs) may use CB (cumulonimbus) to describe this cloud. Appendix A contains selected weather report and forecast contractions.

Clouds with vertical development show up well on visible imagery. They tend to be thick, with large thunderstorms the highest on the list of reflectivity. IR imagery is a good indicator of vertical development. The closer to white, the colder, and therefore the higher the tops. Overshooting cirroform tops are also an indicator of development and severe weather potential. High clouds and those with vertical development also tend to cast shadows, especially during morning and afternoon on visible imagery and certain enhanced IR images.

General cloud characteristics are described in Table 4-2.

Of the two basic cloud types—cumuliform and stratiform—both may form at any level: high, middle, or low. Cumuliform clouds are rounded, billowy, and puffy because they are formed in unstable air. Stratiform clouds are flat and sheetlike because they are formed in stable air. The ability to recognize cloud types on satellite imagery can

Table 4-2 Cloud Characteristics

NAME CONTENT (LWC)	CATEGORY	BASES (FT)	HEIGHT	STABILITY	PRECIPITATION*	EXTENT AND LIQUID WATER
Cirrus (curly)	Cirroform	Above 20,000	High	Stable	None	Ice crystals
Cirrostratus	Cirroform	Above 20,000	High	Stable	None	Ice crystals
Cirrocumulus	Cirroform	Above 20,000	High	Unstable	None	Ice crystals
Altostratus (alto = high)	Stratiform	6500 to 20,000	Middle	Stable	RA SN	Widespread—solid and liquid water
Altocumulus	Cumuliform	6500 to 20,000	Middle	Unstable	VIRGA	Limited—solid and liquid water
Stratus (spread out)	Stratiform	Surface to 6500	Low	Stable	DZ	Widespread—high LWC
Stratocumulus	Stratiform	Surface to 6500	Low	Slightly unstable	DZ – RA	Limited—high LWC
Cumulus (heaped up)	Cumuliform	Near surface	Vertical development	Unstable	SHRA GS	Limited—high LWC
Nimbostratus (nimbus = rain)	Stratiform	Surface to 6500	Low	Stable	– RA RA	Widespread—high LWC
Cumulonimbus	Cumuliform	Near surface	Vertical development	Unstable	+TSRA +RA GR	Limited—high LWC

*See Appendix for abbreviations.

give a pilot a good idea of atmosphere stability and associated aviation weather hazards.

Aviation hazards associated with stable conditions are quite different from those associated with unstable conditions. In general, unstable conditions are associated with thunderstorms, showery precipitation, gusty winds, low-level wind shear (LLWS), convective turbulence, clear icing, good visibility, and an absence of low ceilings. In unstable air, uneven heating causes convective currents and thermal turbulence. Pockets or parcels of warm air rise and create updrafts. The rising air cools and clouds sometime form in the updrafts. Between the updrafts, the air sinks, resulting in an area of clear skies. Stable conditions are usually associated with low ceilings, poor visibility, steady precipitation, rime icing, calm or steady wind, and the lack of convective turbulence. However, mountain-wave and mechanical turbulence can be associated with a stable air mass.

Moisture, Vertical Motion, and Stability

Most aviation weather hazards can be related in terms of moisture, vertical motion, and stability. Moisture is the first player in aviation weather. In fact, without moisture we would essentially have no weather at all. The atmosphere can be moist or dry. Vertical motion is the second player in aviation weather. Upward vertical motion increases relative humidity, clouds, and precipitation. Downward vertical motion decreases relative humidity, evaporates clouds, and precludes precipitation. Vertical motion can result in thunderstorms or clear skies. Atmospheric stability is the third player in aviation weather. Stability determines whether the weather is benign or severe, the ride smooth or turbulent, visibility good or poor.

Vertical motion can be produced, enhanced, or dampened by one or all of the following:

- Convergence and divergence
- Frontal lift
- Dry line
- Vorticity

- Upslope

- Pressure systems

- Convection

- Warm- and cold-air advection

- Thunderstorms

During the following discussion it may be useful to refer to Table 4-3, "Vertical Motion Producers." The table indicates whether motion is upward (destabilizing) or downward (stabilizing). The table also includes those weather products where these phenomena are observed and forecast.

Convergence refers to an inflow or squeezing of the air. On a horizontal surface, when airflow into an area is greater than the outflow, the air literally piles up. Since the ground prevents the air from going downward, there is only one way left for it to go—up. The bottom line: An area of convergence is an area of rising air. Convergence can occur aloft over dense, cold air and is not necessarily confined to a layer bounded by the surface. When moisture is adequate and convergence is great enough, condensation occurs.

Convergence occurs along surface low-pressure troughs and at the center of low-pressure areas. An area of strong winds blowing into an area of lighter winds causes wind speed convergence. When this kind of convergence occurs near the surface, the result is a region of rising air. Clouds caused by convergence appear on satellite imagery.

Divergence is the opposite of convergence. Downward motion (subsidence) of air causes it to spread out at the Earth's surface. Divergence is a drying and stabilizing process. Like convergence, divergence may occur in a layer aloft not extending to the ground. Divergence occurs along high-pressure ridges and at the center of high-pressure areas. When air near the surface blows from an area of light winds into an area of stronger winds, wind speed divergence results. Divergence typically results in clear skies on satellite imagery.

Observed and forecast locations of convergence and divergence can be obtained from the Surface Analysis Chart, 850-mb and 700-mb

Table 4-3 Vertical Motion Producers

PHENOMENON	STABILITY	OBSERVED	FORECAST
Convergence	Up	Surface, 850 mb/700 mb*	SIG WX Prog[†]
Divergence	Down	Surface, 850 mb/700 mb	SIG WX Prog
Fronts	Up	Surface	SIG WX Prog
Dry line	Up	Surface	
Vorticity	Up/Down	500-mb heights/vorticity	500-mb Prog heights/vorticity
Orographic	Up/Down	Surface	SIG WX Prog
High pressure (surface)	Down	Surface	SIG WX Prog
Low pressure (surface)	Up	Surface	SIG WX Prog
Ridge (surface)	Down	Surface	SIG WX Prog
Trough (surface)	Up	Surface	SIG WX Prog
High pressure (aloft)	Down	Constant-Pressure Charts	Constant-Pressure Progs[‡]
Low pressure (aloft)	Up	Constant-Pressure Charts	Constant-Pressure Progs
Ridge (aloft)	Down	Constant-Pressure Charts	Constant-Pressure Progs
Trough (aloft)	Up	Constant-Pressure Charts	Constant-Pressure Progs
Convection	Up		
Warm-air advection (surface)	Up	850-mb, 700-mb Charts	850-mb, 700-mb Progs
Cold-air advection (surface)	Down	850-mb, 700-mb Charts	850-mb 700-mb Progs
Warm-air advection (aloft)	Up	300-mb, 200-mb Charts	300-mb, 200-mb Progs
Cold-air advection (aloft)	Down	300-mb, 200-mb Charts	300-mb, 200-mb Progs
Thunderstorms	Up	Radar Summary	SIG WX Prog/SVR WX Outlook[¶]

*850-mb and 700-mb Constant-Pressure Charts.

[†]Significant Weather Prognostic Chart.

[‡]Constant-Pressure Prognostic Chart.

[¶]Severe Weather Outlook Chart.

Constant-Pressure Charts, and the Low-Level Significant Weather Prognostic (Prog) Chart (SIG WX Prog).

Air masses of different properties (temperature and moisture) do not tend to mix. Differences in temperature, humidity, and wind may change rapidly over short distances. Where there are temperature and moisture differences, there is a difference in density. This zone of rapid change separating the two air masses is a frontal zone, more commonly referred to as a front. In these zones, the less dense air is lifted. This causes vertical motion in the atmosphere. The type of weather produced depends on the stability of the atmosphere. Active fronts are easily recognized on satellite imagery. However, if the front is weak there may be a complete absence of clouds.

A dry line, or temperature–dew point front, marks the boundary between moist, warm air from the Gulf of Mexico and dry, hot air from the southwestern United States. Dry lines usually develop in New Mexico, Texas, and Oklahoma during the summer months. Since the moist air from the Gulf is less dense than the dry, hot desert air, it is forced aloft. If the air mass is unstable, thunderstorms and tornadoes develop along the boundary. As shown in Table 4-3, dry lines appear on the surface analysis chart, and are occasionally forecast on prog charts.

Anything that spins has vorticity, which includes the Earth. *Vorticity* is a mathematical term that refers to the tendency of the air to spin; the faster that air spins, the greater its vorticity. (Note from Table 4-3 that vorticity can produce either upward (unstable) or downward (stable) vertical motion.) A parcel of air that spins counterclockwise (cyclonically) has positive vorticity; a parcel of air that spins clockwise (anticyclonically) has negative vorticity.

Air moving through a ridge, spinning clockwise, gains anticyclonic relative vorticity. Air moving through a trough, spinning counterclockwise, gains cyclonic relative vorticity; therefore, there tends to be downward vertical motion in ridge-to-trough flow, and upward vertical motion in trough-to-ridge flow.

The observed location and forecast position of positive and negative vorticity can be found on 500-mb Heights/Vorticity Charts.

Orographic is a term used to describe the effects caused by terrain, especially mountains. An orographic effect is upslope and downslope. Air can take on the characteristics of the terrain through the process of conduction, as well as through the adiabatic process. That is, the air can absorb heat or moisture by direct contact with the surface. Air moving up a slope rises and tends to cool; air moving down a slope sinks and tends to warm.

Like vorticity, orographic effects can be either upward or downward. Its effects can be inferred from the Surface Analysis Chart for current conditions and the Significant Weather Prog for forecast events, with a knowledge of terrain. In the absence of higher cloud cover, upslope conditions are easily recognized on visible satellite imagery.

Circulation around high-pressure areas produces downward vertical motion, and around low-pressure areas produces upward vertical motion. Typically, but not always, high pressure means good weather and low pressure means poor weather.

Downward vertical motion occurs along high-pressure ridges (commonly referred to as just "ridges"). Conversely, upward vertical motion occurs along low-pressure troughs (usually simply referred to as "troughs").

Vertical motion also occurs at higher levels. For example, air moving from an upper ridge to trough produces downward vertical motion; air moving from an upper-level trough to ridge produces upward vertical motion. The location and forecast position of upper-level ridges and troughs can be found on Constant Pressure Charts and Progs.

Atmospheric convection is the transport of a property vertically. For our purposes, near-the-surface convection is caused by surface heating. Surface heating, and the resulting convection, is a primary vertical motion producer. Areas of convection cannot be directly found on standard aviation weather charts and progs. However, areas of convection can be inferred in areas of high temperatures, especially during afternoon heating. Afternoon cumulus shows up well on visual satellite imagery. However, with unstable conditions, thunderstorms can develop. Because air mass thunderstorms tend to produce an extensive cirrus shield, their exact location may be difficult to determine by satellite imagery alone. By comparing real-time satellite and radar images, the exact location of thunderstorms can be determined.

Advection is a term used to describe the movement of an atmospheric property from one region to another. Temperature, moisture, and stability are properties that can be advected.

From the surface to about 10,000 ft, warm-air advection produces upward vertical motion and cold-air advection produces downward vertical motion. As warmer, less dense, air moves into an area, it will tend to rise. Warm-air advection causes surface pressures to fall. This results in convergence and upward vertical motion. When cooler, more dense, air moves into an area it tends to sink. Cold-air advection causes surface pressures to rise. This results in divergence and downward vertical motion. Therefore, warm-air advection destabilizes conditions, whereas cold-air advection tends to stabilize the weather at and near the surface. Areas of observed and forecast warm- and cold-air advection can be found on 850-mb and 700-mb Constant-Pressure Charts and Progs.

Above the 500-mb level the opposite occurs. Cold-air advection destabilizes conditions and warm-air advection stabilizes the atmosphere. Cold-air advection above the 500-mb level decreases the lapse rate. This enhances any convective activity that might develop. Conversely, warm-air advection aloft stabilizes the atmosphere by increasing the lapse rate; thus retarding any convection. Areas of observed and forecast warm- and cold-air advection can be found on 300-mb and 200-mb Constant-Pressure Charts and Progs.

A tremendous vertical motion producer, a thunderstorm is a local storm produced by cumulonimbus clouds. The storm itself may be a single cumulonimbus cloud or cell, a cluster of cells, or a line of cumulonimbus clouds that in some cases may extend for several hundred miles. Thunderstorms are always associated with one of the previously mentioned vertical motion producers. Observed thunderstorm activity is found on real-time radar displays and the Radar Summary Chart; forecast locations are depicted on the Significant Weather Prog and Severe Weather Outlook Chart, and inferred from the moisture/stability charts.

Stability is the tendency of an air mass to remain in equilibrium— its ability to resist displacement from its initial position. If we move a parcel of air, then remove the lifting mechanism, one of several things occurs: The parcel will tend to return to its original position; the parcel will continue to rise without any additional lifting force; the parcel initially resists upward displacement to a certain point, where it then spontaneously continues upward; or the parcel will remain at the level where the external force ceases. (A parcel is a small volume of air; it retains its composition and does not mix with the surrounding air.) These four processes are respectively known as:

- Absolute stability
- Absolute instability
- Conditional instability
- Neutral stability

A parcel is absolutely stable when it resists vertical displacement, whether saturated or unsaturated. The parcel is always cooler (denser) than the surrounding air. It wants to sink. Thus, vertical motion is impossible, unless caused by an external force.

Air is absolutely unstable when vertical displacement of a parcel within the layer is spontaneous, whether saturated or unsaturated. If a parcel is lifted, its temperature is always warmer than the surrounding air. The cooler, more dense air surrounding the parcel forces it upward. Vertical motion is spontaneous and the layer is absolutely unstable.

Let's look at the third case, conditional instability. Conditional instability refers to the structure of a column of air which will produce free convection of a parcel as a result of its becoming saturated when forced upward. Free convection means that once saturation occurs, upward movement will continue spontaneously. This is the level of free convection (LFC). A lifted parcel is stable to the point where saturation occurs. Below the lifted condensation level, the parcel exhibits absolute stability. However, upon saturation, upward displacement becomes spontaneous. The parcel becomes unstable on the condition that it reaches saturation. High moisture content in low levels and dry air aloft favor instability. Conversely, dry air in low levels and high moisture content aloft favor stability.

The concepts of absolute stability and absolute instability are relatively straightforward. Conditional stability is more complex and includes many subclassifications. It depends not only on temperature, but also on water vapor distribution.

When a parcel is displaced, and remains at rest when the displacing force ceases, the layer is neutrally stable.

Keep in mind that the atmosphere can be dry and stable or moist and stable, as well as dry and unstable or moist and unstable. The weather produced by an air mass is a combination of moisture, vertical motion, and stability. Like vertical motion and moisture, the Area Forecasts or TWEB Route Synopses often provide a clue to atmosphere stability. Other sources are the outlook portion of Convective SIGMETs and the Convective Outlook discussion.

Putting It Together

Refer to Fig. 4-7. A pilot's first task when using a satellite image, or for that matter any aviation weather product, is to determine the date and time of the observation. It doesn't make any sense to use a picture that is hours or even days old. Notice on the top margin of Fig. 4-7 that this image was taken on November 12, 2001 (12 NOV 01) at 2200 UTC (Time = 22:00 UTC). Also note that the resolution is 4 kilometers (km) or approximately 2 nm (RES = 4 KM). Cloud elements smaller than 2 nm will not appear on the image. The other data in the margin identifies the satellite and sector of the observation.

FACT

The time standard used on aviation weather products is Coordinated Universal Time (UTC), at times—pardon the pun— referred to as ZULU or "Z."

It seems that an advisory committee of the International Telecommunication Union in 1970 was tasked with replacing the international time standard of Greenwich Mean Time (GMT). The question became whether to use English or French word order for Coordinated Universal Time—sound familiar? So, instead of CUT or TUC, UTC was adopted and became effective in the late 1980s.

Note the grid in Fig. 4-7. Those portions of the ocean that can be seen correlate exactly with the coast line. This is a fall, afternoon picture from

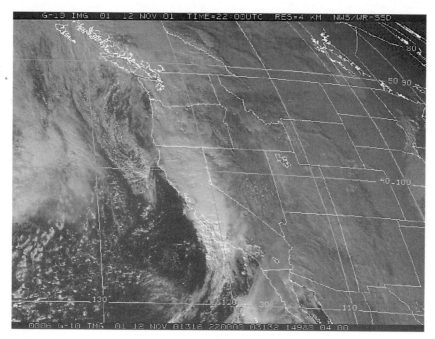

Fig. 4-7 A pilot's first task when using a satellite image, or for that matter any aviation weather product, is to determine the date and time of the observation.

GOES West and the terminator can be seen in the upper right portion of the image. Thick clouds (bright) can be seen along California's coast. This weather was due to a cold front moving through the area. Notice how distinct the frontal boundary appears on the image. Thin clouds (light-gray shaded) cover the southern California deserts and northern Baja. The contrast between the ocean and land is easily seen.

Now refer to Fig. 4-8. This is an enhanced infrared image also taken on November 12, 2001 at 2215 UTC, with a resolution of 4 km. We know that it is enhanced by noting the gray scale at the bottom of the image, just above the satellite data in the bottom margin. Beginning on the right of the gray scale is the number 40, indicating that black is $+40°$C. To the right numbers decrease through $-20°$C which is white on the image. Although it can't be seen, since this image is published in black and white, the image is color-enhanced for temperatures between $-30°$C and $-60°$C. On the color image -30 is yellow, -40 red, -50 blue, and -60 green. But even on this black-and-white reproduction, contouring can easily be seen.

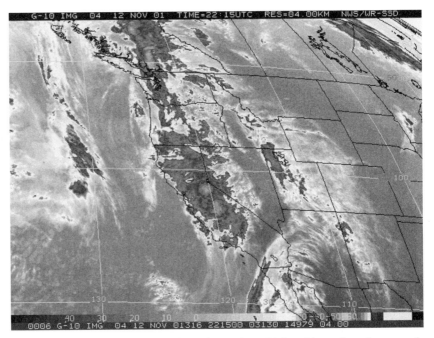

Fig. 4-8 We know that this satellite picture is an enhanced infrared image from the gray scale at the bottom of the image, just above the satellite data in the bottom margin.

The enhancement curve in Fig. 4-8 is designed to contour thunderstorm tops. Contouring typically begins at about $-30°$C, with thunderstorm tops normally in the range of $-40°$C to $-50°$C and overshooting thunderstorm tops at about $-60°$C. (Overshooting tops refer to protuberances from the top of cumulonimbus clouds that rise above the general upper level of the cloud.) In Fig. 4-8 it appears there are two thunderstorm cells with very cold cirrus shields in central California—the two large light areas completely surrounded by darker contouring. These tops have temperatures in the $-50°$C range. Be careful. Cold tops do not necessarily mean thunderstorms. In this example there were some isolated thunderstorms, but most of the activity was nonconvective. How do we know? By checking the Radar Summary Chart.

The 1515Z Radar Summary Chart for November 12, 2001, is shown in Fig. 4-9. Although this chart displays data that are 7 hours older than the satellite images, it depicts an area of rain showers (RW) and thunderstorms (TRW) in central California. The cells are moving from

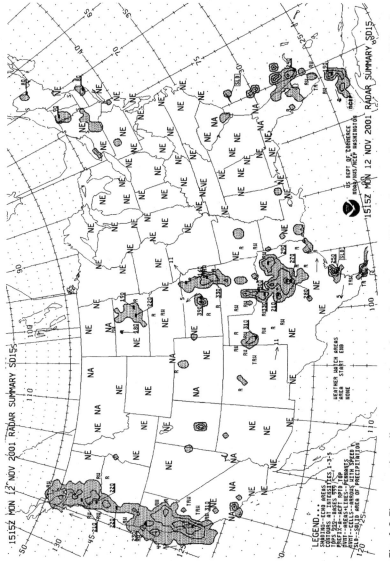

Fig. 4-9 The convective cells are moving from the south-southeast at 44 knots; however, the area movement—which unfortunately is no longer displayed on this chart—is from the northwest. This can be determined from the satellite images.

the south-southeast at 44 knots. However, the area movement—which unfortunately is no longer displayed on this chart—is from the northwest. This movement is easily identified from the satellite images. The radar chart confirms our identification of thunderstorms in central California.

What else could cold tops be? Recall that in Fig. 4-7 thin clouds appeared over the southern California deserts and northern Baja. The IR image indicates these clouds are very cold—white, $-20°$C, and some contouring at less than $-30°$C. These are in fact thin cirrus clouds. Also recall that thin clouds may be depicted at a lower temperature, and therefore, are depicted at a warmer temperature than their actual temperature. This discussion brings up a major point about weather satellite interpretation: Most often, both the visible and infrared images together are needed for correct interpretation. Thick clouds on the visible image and cold tops on the IR image usually mean nimbostratus or cumulonimbus clouds. Thin clouds on the visible image and cold tops on the IR image usually mean cirrus.

Now refer to Fig. 4-10. This is a 1-km-resolution image of southern California for the same date and time as Figs. 4-7 and 4-8. Note the additional detail available on the lower-resolution image. Texturing and shadows can easily been seen with the convective activity north and west of Los Angeles. The bright, white tops indicate the location of possible thunderstorms. *Possible* thunderstorms? Yes. Without additional data, specifically radar, we can make assumptions about the weather from satellite imagery, but to confirm the exact type of weather, and therefore any hazards, other products must be used. From this image alone the only positive conclusion we can make are that the clouds off the coast are thick-white, and the clouds in the desert are thin-gray.

Just as with weather chart interpretation, best results from satellite imagery are obtained by comparing the visible with the infrared image at the same time frame. Often what may be misinterpreted or ambiguous on one image can be resolved by comparing it with the other image. Unfortunately, this doesn't work at night, when only the IR image is available. Optimum interpretation results from a comparison of chart and satellite information—part of the "complete picture."

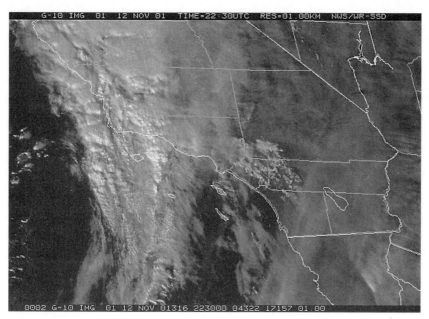

G-10 IMG 01 12 NOV 01 TIME=22:30UTC RES=01.00KM NWS/WR-SSD

0002 G-10 IMG 01 12 NOV 01316 223000 04322 17157 01.00

Fig. 4-10 Texturing and shadows can easily been seen on this high-resolution visible image north and west of Los Angeles.

Now if we have the capability to view several images in succession—a satellite loop—we can often get a sense of weather movement, development, and dissipation. Viewing a loop will also help distinguish between terrain and surface features on one hand and cloud cover on the other.

In subsequent chapters we will discuss the identification of geographical and weather features, and the application of satellite imagery. We will apply the principles of geographical and weather feature identification, and include practical examples of application. Various satellite terms will be defined. Pilots can expect to see this terminology on various weather products, such as Center Weather Advisories, Alert Weather Watches, and the Convective Outlook.

CHAPTER 5

Geographical and Weather Features on Satellite Imagery

In this chapter we will build on many of the subjects discussed in Chap. 4. In order to limit the total number of illustrations, an individual figure may be referred to more than once and occasionally out of order. For the same reason, geographical and weather features will be discussed throughout the chapter.

As mentioned in Chap. 4, typically a distinct boundary occurs between land masses and oceans on both visible and IR satellite imagery. Water surfaces are the least reflective on visible imagery, and there is often a distinguishable temperature difference on IR images, especially during maximum daytime heating. On IR images the gray shade contrast seen at a land-water boundary often appears to reverse at night. Large lakes, bays, and rivers usually can be identified.

Major mountain ranges and valleys also can be seen. Here again, these features usually have different reflectivity and temperature ranges. Deserts are also distinguishable because of the low reflectivity of sand compared to adjacent wooded areas and mountains, and temperature contrasts, especially during maximum daytime heating.

Snow-covered terrain usually appears whiter in the IR imagery than its surroundings. The brightness of a snow area depends, among other things, on whether there is vegetation within the area, on the type of vegetation, and on how much of the vegetation is covered by snow. Light snow on the tops of high mountain ranges is less detectable on IR than in a visible image because the temperature difference between land and snow is small, while the difference in reflectivity between snow and land is large.

It may be difficult to distinguish snow from clouds. On IR images, the clouds can be warmer than snow-covered terrain and appear as darker areas, or they may be the same temperature or colder than the terrain and appear as bright as, or brighter than, the snow-covered areas. With few exceptions, clouds seldom persist in one location for more than a few hours. Therefore, snow fields can be identified by comparing successive images—a satellite loop.

The most dramatic changes in IR imagery are produced by the daytime-nighttime variation of land surface temperature, especially in desert regions.

As mentioned, most satellite imagery contains a grid with cultural boundaries, such as county, state, and international borders; and they often depict large lakes for reference. A knowledge of terrain features within these boundaries is very helpful in determining the location of specific terrain features. Since Visual Flight Rule (VFR) aeronautical charts provide the location of various features, these charts can assist in the location of various cultural and geographical locations.

Figures 5-1 and 5-2 are visible and IR images, respectively, from GOES West. These images are late winter, midday observations (March 2, 1999, 1900Z). Major mountain ranges and valleys are clearly distinguishable, even at this resolution (8 km). Note California's Central Valley, the deserts of southern California, and the dendritic patterns over the Sierra Nevada Mountains. The contrast between land (light) and water (dark) is clearly distinguishable on the visible image. In Fig. 5-2, the IR image, the opposite is true. The warmer land areas are dark and the cooler water is light. Now take a look at the white spot in southern New Mexico, just north of El Paso in Fig. 5-1. You guessed it! It's White Sands.

The main feature on these images is the well-developed comma cloud off the Pacific Northwest. The low-pressure area at the center of the spiral is clearly visible. From the visible image, thick clouds are associated with the comma cloud. Distinct texturing is occurring in the region of the occluded and warm front. The exact location of these fronts is covered by a cirrus shield. However, the cold front position is

Fig. 5-1 Major mountain ranges and valleys are often clearly distinguishable on visible imagery.

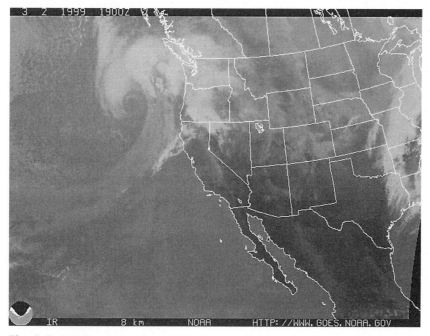

Fig. 5-2 Cloud tops often reveal the strength of weather systems, and can be determined from infrared (IR) imagery.

distinctly visible, along with its surface frontal boundary. Ahead of the front are typical cloud formations associated with this phenomenon. There are considerable, relatively thick clouds along the front in the Pacific Ocean. However, south of the latitude of San Francisco, the front is weak, as revealed by the low tops (gray on the IR image) in this region.

Behind the front is an area of open cell stratocumulus. How do we know these are stratocumulus? The IR image shows dark, warm, low tops. Open and closed cellular patterns are the most common cloud formations found on satellite images. Recognition and interpretation of these cellular cloud patterns enables us to identify regions of cold air advection and areas of cyclonic, anticyclonic, and divergent flow in the cold air behind polar fronts over oceanic areas. Cellular cloud patterns also aid in the identification of cloud types, the location of the jet stream, and regions of positive vorticity advection. They also provide indications of surface wind direction and speed, and atmospheric stability.

These cloud patterns form as a result of mesoscale convective mixing within the large-scale flow. As the cold air to the rear of a cold front is advected over warmer water, the air mass is heated from below. The open cells are composed of cloudless, or less cloudy, centers surrounded by cloud walls. The closed cells are characterized by approximately polygonal cloud-covered areas bounded by clear or less-cloud walls. Open cells form where there is a large air-sea temperature difference and closed cells form where a weaker air-sea temperature contrast exists. The vertical motion from this process is capped by a subsidence inversion away from more active cyclonic areas. The final result is an atmospheric balance of upward and downward vertical motion in the lower levels, resulting in a cellular pattern.

In Fig. 5-1 a large area of the eastern Pacific, behind the front, is covered by cellular patterns formed as cold air moved over the warmer ocean surface. These areas will be characterized by relatively low tops, turbulence below the clouds, and generally smooth air above the clouds to below the flight levels (18,000 ft). In the flight levels, clear air turbulence (CAT) would be a strong probability.

Note the clouds along the southern California and northern Baja coasts. By comparing the visible and IR image we can determine they are coastal stratus. From the visible image, the clouds are relatively thick; however, the IR image reveals their tops have temperatures about the same as sea surface temperatures. Therefore, their tops are very low.

Figures 5-3 and 5-4 are GOES East visible and infrared observations, respectively, for March 10, 1999, at 1815Z.

The upper midwest, especially Wisconsin, Illinois, and Indiana, is snow covered. Note the area south of Green Bay and Lake Winnebago in Wisconsin, and the Illinois River. Even at this resolution, these land features, along with the Great Lakes, help confirm that this is a region of snow, rather than clouds. Additionally, the Surface Analysis and Weather Depiction Charts can be used to confirm that this is a cloud-free area, and that the image, in fact, depicts snow.

Figure 5-3 shows relatively bright clouds throughout the western portion of the Mississippi Valley, central Texas, and the Ohio River

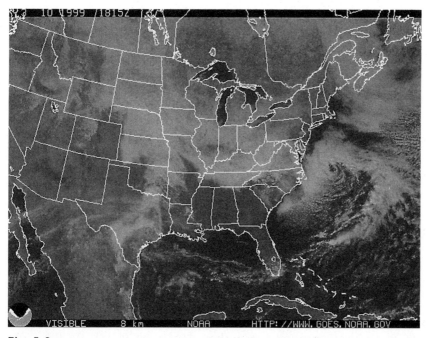

Fig. 5-3 When using any satellite product a pilot should immediately identify the kind of image, date, and time.

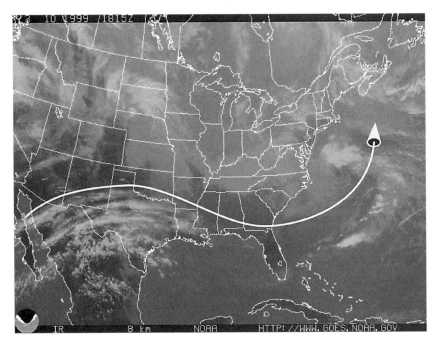

Fig. 5-4 Recall that on IR images, high clouds are cold and bright, low clouds warm and gray.

Valley. What are the cloud bases? Cloud bases for this system cannot be determined from the satellite image. For that information we would have to consult METARs, and the Surface Analysis and Weather Depiction Charts. A disorganized storm system is over the Atlantic. It still retains its more or less comma cloud appearance, with some texturing. To determine precipitation and precipitation intensity we would have to consult radar observations or the Radar Summary Chart.

Over northern Mexico, western Texas, and the Gulf of Mexico are gray to white cloud formations. From this information alone, the only deduction that can be made is that the bright clouds are relatively thick, darker clouds relatively thin. From the IR image in Fig. 5-4, the clouds are bright, indicating high, cold tops. Therefore, we can deduce that these are cirrus. The location of the jet stream is depicted in Fig. 5-4, which will be discussed in detail in a subsequent section.

To continue the analysis, refer to Fig. 5-4 and note that this is an IR image of the same time frame as the visible image in Fig. 5-3. In the vicinity of the Great Lakes, surface and water temperature are the same, as indicated by the same shade of gray. From the IR image alone its difficult to distinguish the snow cover of the upper midwest from low clouds. However, by using the two images together we can conclude it is indeed snow cover. In Florida and Mexico land temperatures are much warmer than the oceans, as revealed by much darker (warmer) land masses.

The relatively bright area over the Dakotas indicates high, cold tops. For Nebraska, Kansas, and the Ohio River Valley, darker gray represents lower, warmer tops.

WHY DOES LAKE ERIE FREEZE?

Lake Erie often freezes in winter while the other Great Lakes remain ice-free. In the autumn, surface lake temperatures decrease as air temperatures lower. Just like air in the atmosphere, cooler, denser surface water sinks as warmer water from the depths rises. This mixing continues through midwinter when water throughout the lake reaches about 4°C. With further cooling, surface water becomes less dense, which sets the stage for freezing. The other Great Lakes are much deeper and almost never reach 4°C throughout their depths. Upwelling of warmer water prevents their freezing.

What is occurring over Illinois, Indiana, and central portions of Kentucky and Tennessee? The visual image indicates relatively bright reflectivity. The IR image reveals temperatures colder than surrounding clear areas to the west. A review of the Radar Summary Chart for this date and time reveals no precipitation echoes in these areas. From this information alone, it appears this is an area of snow cover.

Cloud cover begins over Ohio and eastern Kentucky and Tennessee. This is revealed by the distinctly colder temperature on the IR image. The radar chart also supports this conclusion with areas of light to moderate rain in these areas. The activity is relatively benign, except in the eastern third of North Carolina and along the central Atlantic coast where thunderstorm activity is reported.

Refer to the activity along the southeast Atlantic coast. There is a narrow, organized line with textured tops indicating thunderstorms with overshooting tops, probably into the lower stratosphere. The cirrus anvils are clearly visible on the visible image, indicating upper-level winds are from the southwest. The IR image shows the highest, coldest tops are along the line depicted in the visible image. The exact location, intensity, and movement of this activity can be determined from the Radar Summary Chart.

Major Weather Systems

For our purposes, major weather systems consist of synoptic scale events, that is, large-scale weather patterns the size of the migratory high- and low-pressure areas and frontal systems of the lower troposphere. Additionally, we will include the jet stream, lake effect, major sea breeze areas, and upper-level low- and high-pressure areas. Finally, we will include the hurricane, which is the best example of a major, nonfrontal weather producing system.

Land and sea breezes and the lake effect can be detected on satellite imagery. Like mountain waves, these phenomena appear on IR imagery, but are typically easier to see on a visible image.

Widespread areas of haze, smoke, dust, sand, and volcanic ash can be seen on visible, and at times IR, imagery. Haze on visible imagery is most easily seen over dark ocean surfaces and rarely shows up well on IR imagery. During the summer, haze boundaries may indicate frontal or air mass boundaries. Haze and smoke are most easily seen on visible imagery during the early morning or late afternoon. Smoke appears as light gray. The NWS uses various enhancements to highlight areas of haze, dust, sand, and volcanic ash for weather advisory and forecast purposes. These enhancements are normally not available to pilots in an operational environment. The NWS has a special Internet site for volcanic ash products at www.ssd.noaa.gov/VAAC/washington.html.

Air Masses

Air masses tend to show up well on satellite imagery. Large high-pressure areas are often cloud free–zones. Low-pressure areas, with moisture present, contain large, organized cloud patterns, the ultimate example being a hurricane. Boundaries between air masses—fronts— again with moisture present, show up well on satellite imagery. Large weather systems often appear in the shape of a comma. A comma cloud indicates an area of low pressure with occluded, warm, and cold fronts, and the jet stream. Maximum vorticity—upward vertical motion—occurs in the center. The surface location of fronts, especially warm and occluded, may be masked by higher clouds in the form a cloud shield—a broad cloud pattern—or overrunning cirrus. Fronts may be indicated by cloud bands, a nearly continuous cloud formation. However, under extremely dry conditions, frontal boundaries may be cloud-free.

In Figs. 5-1 and 5-2, the frontal boundary between the two air masses is clearly distinguishable. Behind the front is a maritime Pacific cold air mass; over the southwest United States is a continental tropical warm air mass. In Figs. 5-3 and 5-4, a ridge of continental tropical warm air exists over the southwest with a trough of maritime tropical warm air off the Atlantic coast.

Fronts

At about 30° latitude are the prevailing westerlies. These are due to the subtropical highs, illustrated by the letter H in Fig. 5-5. At the poles is an area of subsidence. This is the region of the polar high—some of the highest atmospheric pressures ever recorded have occurred in these areas. (In the late 1980s pressures well above 31.00 inHg occurred in Alaska. This precipitated emergency regulations because most aircraft altimeters can be corrected only for pressure up to 31.00 in.) This is also the area of the polar easterlies at about 60° latitude.

The sloping boundary between contrasting air masses is known as a *front*. Between the polar easterlies and the midlatitude westerlies is the polar front. This is an area of global convergence of warm air from the south and cold air from the north. The polar front is more or less continuous around the world as shown in Fig. 5-5. However, where it is weak there may be areas of little or no weather as illustrated in Fig. 5-5 over the western United States and in the central Pacific Ocean.

Frontal boundaries are three dimensional. They extend vertically as well as horizontally, upward over the colder, denser air and exhibit an abrupt temperature difference throughout their vertical extent. The more or less permanent, undulating polar front boundary is known as a baroclinic zone. A baroclinic atmosphere results in a strong, active

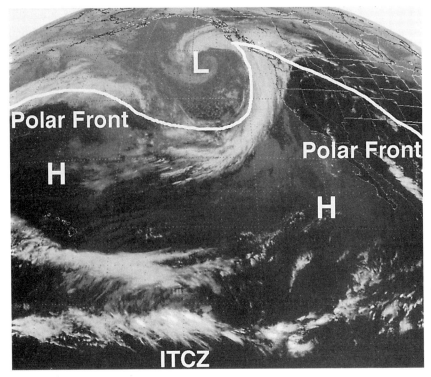

Fig. 5-5 The components of the general circulation are easily identified on satellite imagery.

front. The opposite is a barotropic atmosphere. With a barotropic atmosphere, density differences across the front are small, resulting in a weak, inactive front.

Fronts can be identified on satellite images by their distinctive cloud bands. As a rule the well-defined cloud bands that occur with active cold fronts and occluded fronts occur in zones of strong baroclinicity. Active warm fronts also occur in strong baroclinic zones. However, the exact location of the warm front is often covered by a cirrus shield. The frontal cloud bands generally extend several thousand miles in length and may exceed 300 miles in width. Cloud bands generally consist of multilayered clouds with cirroform cloud cover. In addition to the long cloud bands, fronts can sometimes be identified from the difference in cloud patterns on each side of the frontal boundary.

Active cold fronts have upper air winds parallel, or nearly parallel, to the frontal zone. This results in a broad band of multilayered low, middle, and high clouds. Inactive cloud fronts have upper air winds perpendicular to the frontal zone that result in considerable subsidence over the frontal zone. This reduces the amount of clouds associated with the front.

Active cold fronts are seen as continuous, well-developed cloud bands. The frontal cloud bands are associated with strong baroclinic zones that have considerable thermal advection and strong vertical shear—strong upward vertical motion. The upper-level winds are parallel, or nearly parallel, to an active cold front, and this, along with the strong baroclinic factors, leads to the well-developed cloud bands. The bands are made up of lower stratiform and cumuliform layers, with cirroform clouds over the top. This is illustrated in Figs. 5-1 and 5-2.

Inactive cold fronts are often seen as narrow, fragmented, discontinuous cold bands. These cold fronts over water occasionally have the same appearance as active cold fronts over land. The inactive cold fronts are associated with weak baroclinic zones that suggest weak cold air advection and insignificant vertical shear. The upper-level winds tend to be perpendicular, or nearly perpendicular, to the front. The light winds and weak vertical shear associated with the baroclinic zone are the contributing factors in the fragmented appearance of the cloud band. The band of clouds is composed mainly of low-level cumuliform and stratiform clouds, but some cirroform clouds may be present. Inactive cold fronts over land may have few or no clouds. Recall that this was illustrated in Fig. 5-5.

Like cold fronts, stationary fronts can be either active or inactive. The active stationary front tends to have the upper flow parallel, or nearly parallel, to the frontal zone. Stationary fronts appear as wide cloud bands. Waves frequently develop on such frontal bands. Inactive stationary fronts are usually found in lower latitudes with a generally west-east orientation. The presence of a subtropical high leads to the dissipation of the clouds in the frontal zone. These fronts are seen as fragmented cloud bands, often devoid of low and middle clouds as a result of the subsidence associated with the subtropical high.

The strong vortex of the occluded front may be located near the trailing edge of the cloud band. This is illustrated in Figs. 5-1 and 5-2. The surface location of the cold front is located near the leading edge of the cloud band, with the warm front hidden by the cloud shield.

Active warm fronts are, at best, difficult to locate on satellite imagery. Inactive warm fronts cannot be located at all. An active warm front may be placed somewhere under the bulge of clouds that is associated with the peak of the warm sector of the frontal system. (The warm sector is that area ahead of the cold front and behind the warm front.) The clouds in the bulging portion of a mature frontal system with an active warm front are combinations of stratiform and cumuliform clouds beneath a cirroform shield. Inactive warm fronts usually have very little baroclinicity associated with them and very few clouds.

The occluded front is located in the band of clouds poleward of the bulge of cirrus over the front, associated with that part of the cloud band that curves in a spiral away from the peak of the warm sector toward the associated vortex. The peak of the warm sector is located equatorward of the cirrus shield under the bulge in the frontal band. Sometimes there is a change in cloud character from smooth-appearing cirroform over the bulge of a frontal system to lumpy-appearing cumuliform poleward of the bulge. This change in cloud character helps to position the point of occlusion and the peak of the warm sector. This is illustrated in Figs. 5-1 and 5-2.

Frontal cloud bands usually change character as they move inland after having had an overwater trajectory. A well-developed continuous frontal cloud band over water can change to a discontinuous, fragmented collection of clouds as the associated front moves over land. This change in cloud character is the result of a decrease in the amount of moisture available over the land. Similarly, frontal cloud bands moving offshore change in appearance after a sufficiently long trajectory over land. As a rule, frontal cloud bands tend to have more continuous clouds over water than over land, although there are cases of continuous frontal cloud bands over land.

The Jet Stream

The jet stream was virtually unknown until World War II, when pilots flying at high altitudes reported turbulence and tremendously strong winds. These winds blew from west to east near the top of the troposphere. Not until 1946 was the jet stream fully recognized as a meteorological phenomenon.

Sharp horizontal temperature differences cause strong pressure gradients that result in the jet stream. Temperature changes rapidly with height. The atmosphere compensates for extremely cold air in polar regions with relatively warmer air above. This relatively warmer air above the tropopause extends well up into the stratosphere. In such zones, the slope of constant pressure surfaces increases with height. Since fronts lie in zones of temperature contrast, the jet is closely linked, or associated with frontal boundaries. When wind speed becomes strong enough, the flow is termed a *jet stream.*

A jet stream is a narrow, shallow, meandering area of strong winds embedded in breaks in the tropopause. Two such breaks typically occur in the northern hemisphere: the Polar Jet located around 30° to 60° north latitude at an approximate height of 30,000 ft, associated with the polar front, and the Subtropical Jet around 20° to 30° north latitude at approximately 39,000 ft. To be classified a jet stream, winds must be 50 knots or greater, although, winds generally range between 100 and 150 knots; winds can reach 200 knots along the east coast of North America and Asia in winter when temperature contrasts are greatest.

A "jet" is most frequently found in segments 1000 to 3000 miles long, 100 to 400 miles wide, and 3000 to 7000 feet deep. The strength of the jet stream increases in winter in mid- and high latitudes when temperature contrasts are greatest, and shifts south with the seasonal migration of the polar front.

The presence of jet streams has a significant impact on flight operations. The jet stream can cause a significant head-wind component for westbound flights, increasing fuel consumption and requiring additional landings.

Another factor associated with the jet is wind shear turbulence. Wind shear is caused by a change in wind speed—either horizontal or vertical—or direction. With a significant wind change over a relatively small distance, severe turbulence can result. Maximum jet stream turbulence tends to occur above the jet core and just below the core on the north side. Additional areas of probable turbulence occur where the polar and subtropical jets merge or diverge. With an average depth of 3000 to 7000 ft, a change in altitude of a few thousand feet will often take the aircraft out of the worst turbulence and strongest winds.

The location of the jet stream is frequently identifiable on satellite imagery. The jet stream usually crosses an occluded system just to the north of the point of occlusion. Cirrus is often associated with the jet stream and high-altitude turbulence. Cirrus that forms as transverse lines or cloud trails perpendicular to the jet stream indicates moderate or greater turbulence. These clouds might be reported as cirrocumulus. Cirrus streaks, parallel to the jet, are long, narrow streaks of cirrus frequently seen with jet streams. Typically, the sharp northern edge of a cirrus cloud shield indicates the location of the subtropical jet stream, and the clouds are known as jet stream cirrus.

Cirrus clouds predominate on the equatorial side of the jet stream and in the anticyclonic portion of the jet. The poleward boundary of the cirrus is often very abrupt; it lies under or slightly equatorward from the jet axis, and frequently casts a shadow on the lower cloud that is clearly visible in satellite imagery. This is illustrated in Fig. 5-4 over northern Mexico and Texas.

Over the oceans, where the jet stream curves cyclonically, the differences in stability on each side of the core are reflected in the appearance of the clouds. On the left side, looking downstream, cold temperatures and unstable air occur, resulting in great vertical development of convective clouds in an open cellular pattern. On the right side of the core, subsidence and warmer air occur and stratiform cloudiness appears in closed cellular patterns.

The main jet stream cloud features are long shadow lines, large cirrus shields with sharp boundaries, long cirrus bands, cirrus streaks, and transverse bands within cirrus cloud formations.

Recall the high, cold clouds over Mexico and Texas in Figs. 5-3 and 5-4. These clouds have an anticyclonic curvature. Since they appear dark on the visible image, we can conclude they are high, thin cirrus clouds, and are associated with the jet stream. The jet stream runs just north of the band from central Baja California through southern New Mexico and the Texas panhandle. The clouds over the Gulf of Mexico are dark, indicating they have low, warm tops. This, along with their cellular appearance in the visible image, indicates stratocumulus.

Refer to the storm system off the Atlantic in Figs. 5-3 and 5-4. The IR image shows high, cold tops along the cold front boundary, in the regions of texturing on the visible image. This indicates considerable vertical development, possibly thunderstorms. High, cold, thick clouds also exist in the area of low pressure at the comma's head. Low, thick, warm clouds are present to the southwest of the low center. From the visible image they appear to be closed cell stratocumulus, becoming open-celled to the southeast. Between the stratocumulus and the cold front is an area known as the dry slot. A dry slot is an area of sinking air beneath the jet stream caused by the intrusion of dry, relatively cloud-free air.

From the satellite images in Figs. 5-3 and 5-4, we can conclude there is an upper ridge of high pressure over the southwest United States, with an upper trough of low pressure along the southeast Atlantic coast. Conditions are relatively clear in the southern tier of states in the area of ridge-to-trough flow. However, the area of trough-to-ridge flow supports the weather system off the Atlantic coast. The jet stream runs from the Texas panhandle, then weakens over an area of relatively high pressure at the surface in the southeast United States, then curves northeast through the dry slot in the comma cloud over the Atlantic, and exits northeast over the Atlantic Ocean.

Lake Effect

Often in winter, cold air moves over relatively warm lakes. The warm water adds both heat and water vapor to the air. The added heat makes the air unstable, resulting in showers to the lee of the lakes. This is known as lake effect. Since it's winter, snow showers develop downwind. These snow showers can be heavy and produce severe aircraft icing. This often occurs in the Great Salt Lake area of Utah, over the Great Lakes, and to a lesser degree over smaller lakes. In November 1996, severe lake effect caused heavy snow and the closing of Cleveland's Hopkins International airport for days. During the period, several aircraft slid off the runway.

CASE STUDY

During the first week of January 1988, lake effect dumped 70 inches of snow on the Tug Hill region of New York state. Areas only a few miles north and south experienced less than 6 inches of snow!

Figure 5-6 contains midday satellite images, with the visible image on the left, and a same-time-frame IR image on the right. On this day, lake-effect snow was falling over and downwind of the Great Lakes, considerable cloud cover existed over the northeast, and a line of thunderstorms had developed over the Atlantic, east of the Florida, Georgia, and South Carolina coast.

The visible image shows thick cloud cover over the Great Lakes. Because of a northwesterly wind, there is a clear area along the northern coastline of Lake Superior, and to a lesser degree the western coastline of Lake Michigan. The IR image reveals a distinct temperature—gray shade—difference between the surface water temperatures of Lake Superior and the cloud layer. This indicates that there is some height to the clouds. We would expect the greatest snowfall to the southeast of the lakes. This could be confirmed by surface observations and the Radar Summary Chart. In fact, this is a weak event, as illustrated by the Radar Summary Chart in Fig. 5-7. The radar chart shows an area of snow falling along the southern shore of Lake Michigan.

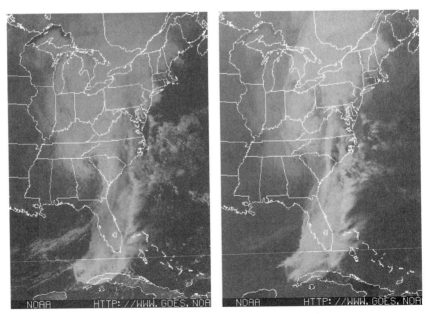

Fig. 5-6 Often a more complete picture can be obtained by comparing visible and IR imagery.

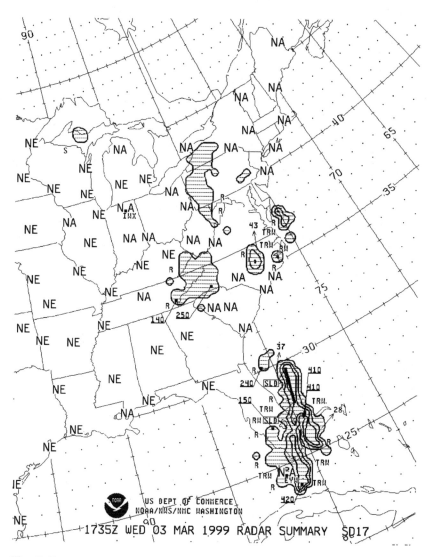

Fig. 5-7 The Radar Summary Chart can usually confirm the intensity and movement of weather depicted on satellite imagery.

Sea Breeze

During the day the land heats, but the water temperature remains relatively the same. The land heats the air near the surface through the process of conduction. The air warms, becomes less dense, and rises. The cooler, more dense air over the water moves into the relatively lower-pressure area onshore. Since the wind blows from the sea to the land, it is called a *sea breeze.*

With enough moisture and lift, clouds develop at the lifted condensation level (LCL) over the land. This is particularly true in the southeast United States, with its abundant moisture and unstable air. Activity tends to be widespread, with some areas receiving torrential downpours, while adjacent areas remain dry. Sea breezes, since they occur during the day, can often be seen on visual satellite imagery once clouds develop.

At night this circulation is reversed. The land cools more rapidly than the sea, resulting in subsidence. The wind blows from the cool land toward the warmer water, creating a *land breeze.* Here again, if moisture, lift, and instability are right, thunderstorms and rain showers develop.

The effect of the sea breeze appears on satellite imagery as a line of clouds—the sea breeze front—inland from the coast with a relatively cloud-free region along and off the coastline. The sea breeze front may or may not be continuous, depending on terrain and other factors. The clear area ends abruptly over water, where either random cloudiness or another line of clouds appears. The vertical development of this secondary line of clouds over the water in most cases is less than the vertical development of the line of clouds inland from the coast.

Where abundant moisture is available along with unstable air, thunderstorms develop. These conditions also occur with large lakes. Refer to Fig. 5-8. The top image was observed at 1515Z, or 8:15 A.M. Eastern Daylight Time (EDT). Both land and sea areas are relatively cloud-free. As the land heats during the day, convection begins. Cool air over the water rushes in to replace the rising air. The middle image of Fig. 5-8 occurred at 1815Z, or 2:15 P.M. EDT. Note that cumulus clouds have developed over most of Florida, with cumulonimbus occurring along the coast, especially east of Lake Okeechobee. The bottom image in Fig. 5-8 occurred at 2015Z, or 4:15 P.M. EDT. Thunderstorms have developed over the southern three-quarters of the Florida land mass. This is an excellent example of weather development as seen on satellite imagery.

Fig. 5-8 A sea breeze is often depicted on visible imagery as scattered cumulus over land areas, with relatively clear skies over the water.

Upper-Level Lows and Highs

Like low- and high-pressure areas at the surface, upper-level lows and highs also occur. Their definition is the same. An upper-level low is an area completely surrounded by higher pressure, and an upper high is an area surrounded by lower pressure.

At and above the 500-mb level (approximately 18,000 ft) when an area of low pressure is completely surrounded by a contour, it is called a *closed low* or *cutoff low*. These cutoff lows, sometimes referred to as *cold lows aloft* or *upper level lows,* are an important winter weather feature. Under these conditions low pressure is reflected from the surface to the tropopause. When the area of low pressure is vertical through the atmosphere, these storms tend to be powerful and erratic. Storms resulting from cutoff lows produce precipitation and low ceilings and visibilities over widespread areas. Forecasting the formation of cutoff lows, and their movement, is difficult. The cloud systems that develop with these storms may have no history of having been associated with a front; however, they often contain cloud bands that appear similar to a front.

How do cutoff lows form? Typically, an intense high-pressure ridge is present over the eastern Pacific, with strong northerly or northwesterly winds aloft along the coast of the Pacific Northwest, with its associated jet stream. The jet typically works around to the south side of the trough, and finally to the east side. After formation, the lows tend to move southward for the first 12 to 24 hours, in response to the strong northerly jet on the west side of the low. Cutoff lows tend to be slow-moving and erratic. Under their influence, weather can remain poor for several days or more. When the low develops over the Great Basin of Nevada and Utah, it is often called an *Ely low.* When these lows move over the north-central Rockies and plains, widespread blizzard conditions and heavy snow results. During summer months, closed lows aloft support the development of thunderstorms, once surface lifting begins.

The following Area Forecasts synopsis illustrates the impact of an upper-level weather system:

STRONG UPPER LOW OVER SOUTHWESTERN NEW MEXICO WILL MOVE TO NORTH CENTRAL TEXAS BY 22Z.

The Surface Analysis Chart showed weak surface high pressure over the western United States. However, the Weather Depiction Chart discloses extensive areas of IFR and marginal VFR, with rain and snow occurring throughout New Mexico, Texas, and Oklahoma. The weather closed airports for days, and was blamed for the deaths of dozens of people. The 500-mb analysis revealed the culprit. A deep upper-level low along the southern Arizona-New Mexico border, and the associated downstream trough produced devastating surface conditions.

CASE STUDY

During one episode, an upper low drifted in the vicinity of Red Bluff, California, for 5 days. Pilots would call day after day wanting to know when the weather would clear. After a while, the common response became, "The low is forecast to move east out of the area tomorrow, but that's what they said yesterday."

Clear air turbulence (CAT) is common in the winter season. The jet stream, lows aloft, and sharp troughs can cause severe CAT. Sharp troughs can produce severe wind shear turbulence as low as 8000 to 10,000 ft.

CAT is implied by the jet stream, lows aloft, and sharp troughs, which can be found on constant-pressure charts. Forecasts for CAT are contained in the SIGMETs and the AIRMET Bulletin.

Upper-level lows tend to form bands of weather, as can be seen in Fig. 5-9. The upper-level low is centered off the central-southern California coast. The front is moving into the southern Sierra Nevada Mountains, southern California, and northern Baja, Mexico. Note the bands of cloud behind the front, associated with the low. The weather deteriorates as a band moves through, then improves, only to deteriorate with the next band. The Area Forecast cannot, and does not, take this into account. Under these

Fig. 5-9 Bands of cloud behind the front, associated with an upper-level low, cause the weather to deteriorate as a band moves through, then improve, only to deteriorate with the next band.

conditions, a pilot must be careful not to get suckered by a temporary improvement.

Upper-level highs typically bring relatively warm, dry, stable weather. This is illustrated by cirroform clouds and otherwise clear skies over the southwest United States in Figs. 5-3 and 5-4 and the H in Fig. 5-5. They tend to block approaching weather systems and are sometimes known as *blocking highs.* Heat waves can develop when the subsidence from an upper-level high lingers over an area. A specific pattern associated with upper-level highs is called an *omega block.* Upper-level flow resembles the Greek letter omega (Ω), with an upper-level high in the center of the pattern. This is a strong blocking high that can remain stationary for days or weeks.

Flying weather in areas dominated by upper-level highs is similar to conditions under the influence of a stable air mass—typically, poor visibilities. High temperatures result in high-density altitude. Hot, dry conditions are ideal for brush and forest fires with their resulting smoke layers. If the wind picks up, pilots will often have to contend with widespread areas of blowing dust and sand.

Hurricanes

Hurricanes produce just about every kind of nasty weather, extending over thousands of square miles. Figure 5-10 is a satellite photo of hurricane Andrew on August, 24, 1992.

Before we proceed, it might be interesting to see how hurricanes get their names. It appears that the practice of giving Atlantic hurricanes women's names began with writer George Stewart in his book *Storm,* published in 1941. A character in the book was a

FACT
Like the numbers of great baseball players, the names of great hurricanes are retired. The World Meteorological Organization retired Andrew's name, making it the 44th western-hemisphere storm name to be so honored (?). Hurricane Andrew just missed category 5 status as it blasted into southern Florida with a central pressure of 27.23 inches.

Fig. 5-10 Hurricanes produce just about every kind of nasty weather, extending over thousands of square miles. The extent of hurricane Andrew can be seen in this satellite image taken on August, 24, 1992.

Weather Bureau meteorologist who used this method as he tracked these storms.

During World War II, Army and Navy weather types tracked storms over the Pacific as part of the war effort. Using women's names to communicate storm information was short, quick, and less confusing than methods used in the past. This practice continued after the War and was applied to Atlantic hurricanes.

Names are alphabetically selected in advance and applied to successive seasonal tropical storms, starting with A and proceeding through the alphabet. Names selected are short, easily pronounced, quickly recognized, and easy to remember. A similar procedure is used for Pacific storms. The only region where tropical cyclones are not named is the north Indian Ocean.

In the 1980s, the practice of using only women's names was changed. Tropical storms are still named alphabetically, with A the first storm of the season, however, male (e.g., Andrew) and female names are used alternately.

Tropical cyclone is a general term applied to any low-pressure area that originates over tropical oceans. (*Cyclone* comes from the Greek *kyklon*, which refers to the coil of a snake. The word *hurricane* does not have a clear origin, but forms of it seem to have been used by the natives of the West Indies and Central America to mean great wind.)

> **FACT**
> What happens if a tropical storm or hurricane moves from the Atlantic into the eastern Pacific? The storm may be the same, but the name changes. In 1988, hurricane Joan underwent a "sex change" as it moved through Nicaragua, ending up as tropical storm Omar in the Pacific.

Tropical storms typically develop during the mid- to late summer season. Principle hurricane months are August, September, and October; most occur in September. Ninety-five percent of the intense hurricanes in the Atlantic occur during this period. However, early season hurricanes can develop during May, June, or July. Hurricanes that affect North America evolve in the warm tropical waters of the Atlantic, Caribbean, and Gulf of Mexico, and off the west coast of Mexico.

Tropical cyclones are classified according to their intensity based on average 1-minute wind speeds. Wind gusts in these storms may be as much as 50 percent greater than the average 1-minute wind speeds. Tropical cyclones are internationally classified as:

- *Tropical depression*—highest sustained winds up to 34 knots

- *Tropical storm*—highest sustained winds 35 to 64 knots

- *Hurricane*—highest sustained winds 65 knots or more

Tropical cyclones develop under optimum sea surface temperature and weather systems that produce low-level convergence and cyclonic wind shear. They favor tropical, easterly waves, troughs aloft, and areas of converging northeast and southeast trade winds along the intertropical convergence zone. [The intertropical convergence zone (ITCZ) is illustrated in Fig. 5-5.]

The low-level convergence associated with these systems, by itself, will not support the development of a tropical cyclone. The system must also have horizontal outflow—divergence—at high troposphere levels. Recall the earlier discussion of high pressure over low pressure. At high levels there is anticyclonic flow. This combination creates a "chimney" in which air is forced upward, causing clouds and precipitation. Condensation releases large quantities of latent heat, which raise the temperature of the system and accelerate upward motion. The increased temperature lowers surface pressure—recall our discussion of surface warm-air advection—which increases low-level convergence. This draws more moisture-laden air into the system. When these chain reaction events continue, a huge vortex is generated that may culminate in hurricane force winds.

Tropical cyclones usually originate between 5° and 20° latitude. Tropical cyclones are unlikely within 5° of the equator because Coriolis force is so small, cyclonic circulation cannot develop. Winds flow directly into an equatorial low, and it rapidly fills.

Tropical cyclones in the northern hemisphere usually move in a direction between west and northwest while in low latitudes. As storms move toward midattitude, they come under the influence of the prevailing westerlies. Thus a storm may move very erratically, reverse course, or even circle. As the prevailing westerlies become dominant, storms recurve toward the north, then to the northeast, and finally to the east-northeast as they reach well into midlatitudes.

If a storm tracks along a coast line or over open sea, it gives up, slowly unleashing its devastation far from tropical regions. However, if the storm moves inland, it weakens as a result of surface friction and loss of its moisture source. Like the process of air mass modification, as storms curve toward the north or east, they begin to lose their tropical characteristics and acquire characteristics more like low-pressure areas in midlatitudes. Strength weakens as cooler air flows into the storm.

While developing, the cyclone has the characteristics of a circular area of broken to overcast, multilayered clouds. Numerous rain showers and thunderstorms are embedded in these clouds. Coverage ranges from scattered to almost solid. Lightning is relatively scarce in hurricanes. This is because updrafts are relatively weak, compared to those in continental thunderstorms.

As cyclonic flow increases, thunderstorms and rain showers form into broken or solid bands paralleling the wind flow that spirals into the center of the storm. These spiral rain bands frequently are seen on radar and satellite images. In both Figs. 5-10 and 5-11, these cloud bands can easily be seen. All the hazardous weather associated with thunderstorms, including tornadoes, is present. Between the bands, weather is less severe, but like upper-level lows, can lure an unwary pilot.

The "eye" usually forms in the tropical storm stage and continues through the hurricane stage. Near the top of the thunderstorms, the air is relatively dry. Loosing its moisture, the air begins to flow outward away from the center, in a diverging anticyclonic flow. This flow extends several hundred miles from the eye. The air begins to sink and warm as it reaches the limit of the storm. This results in clear skies outside the storm.

Surrounding the eye is a wall of clouds that may extend above 50,000 ft. This "wall cloud" contains torrential rains and the strongest winds in the storm. As a result of eye-wall thunderstorms, the air warms from the release of latent heat. This initiates a downward motion in the eye, which helps account for the absence of weather at the storm's center. In the eye, skies are cloud-free, flight conditions smooth, and winds comparatively light. The average diameter of the eye is between 15 and 20 miles, although sometimes smaller or larger.

Figure 5-11 shows two Pacific storms—tropical storm Eugene at 140° W and hurricane Dora at 125° W longitude. Since their energy comes from the ocean, they dissipate rapidly over land. Occasionally, eastern Pacific hurricanes reach Hawaii and southern California. The moist, unstable remnants of these storms can be carried north and inland to affect central California and the southern part of the

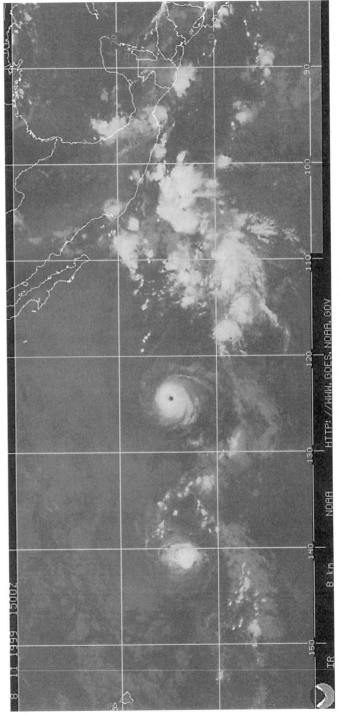

Fig. 5-11 Since their energy comes from the ocean, hurricanes typically dissipate rapidly over land.

intermountain region. In Fig. 5-12, moisture resulting in thunderstorms from the remnants of tropical storm Hillary can be seen as far north as central California, southern Nevada, and Arizona. Analysis of these southerly disturbances was first begun in the 1930s. Because they sometimes approached from the southeast, they were called *Sonoran storms,* after the Mexican state of that name.

Convective Weather and Convective Weather Systems

Convective activity appears very bright on both visible and IR imagery, as a result of thick cloud layers and high, cold tops. Often, individual cells can be identified with an air mass thunderstorm. Organized lines, produced by fronts and squall lines, also can be seen. Once the cells or lines have developed, the exact location of the thunderstorm cells may be masked by overrunning cirrus clouds. However, overshooting tops are most evident on visible imagery with a low sun angle. As air flows rapidly out of the storm at the surface, a line of clouds (an arc cloud) forms along the leading edge. Arc lines (clouds) are a good indicator of severe or extreme turbulence. Outflow boundaries are sometimes depicted on the Surface Analysis Chart. Often, new thunderstorms develop along these outflow boundaries. At a point where two or more arc lines converge, convection is strongest, which can produce severe weather.

Figure 5-13 illustrates a large weather system, with convective cells hidden by the general cloud mass. At the left is an infrared image and at the right is a visible image from the same time frame. Figure 5-13 shows a late winter storm system moving through the southern Plains states and into the southern Mississippi Valley. From the visible image (Fig. 5-13 right), we see that the cloud mass is thick (white); from the IR image (Fig. 5-13 left) the tops are cold (white). At this point the only thing we can conclude is that this is a thick cloud band with high, cold tops.

The images in Fig. 5-13 also illustrate the difference between land and water on satellite imagery. Note the Great Lakes and the Gulf of California. In the visible image, the water is dark and the land is gray. In the IR image the Great Lakes have the same temperature range as

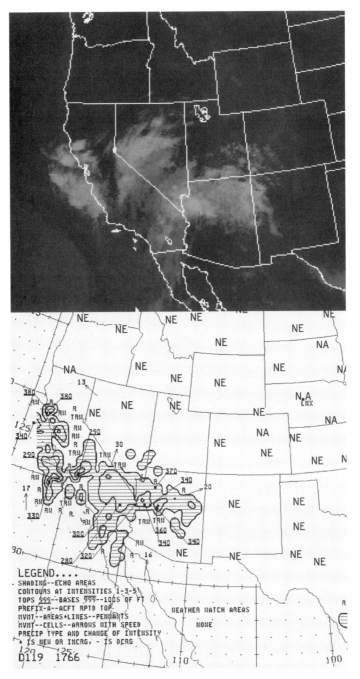

Fig. 5-12 The moist, unstable remnants of Pacific hurricanes can be carried north and inland to affect central California and the southern part of the inter-mountain region.

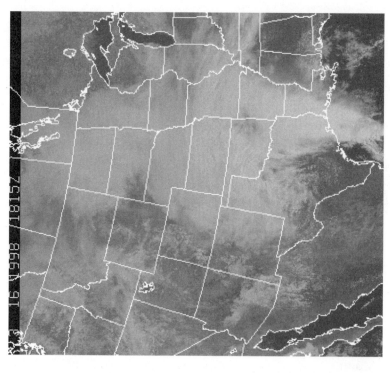

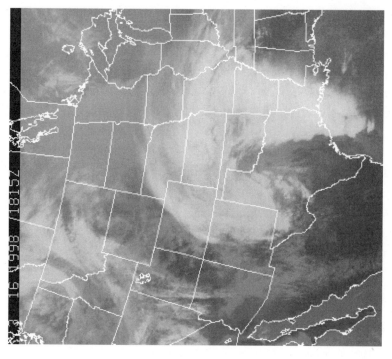

Fig. 5-13 The left shows an infrared image and right a visible image from the same time frame. This is a large weather system, with convective cells hidden by the general cloud mass.

the land, which is what we would expect in late winter. On the other hand, the land mass of Baja California and Mexico is warmer than the adjacent water areas and therefore is darker than the ocean waters.

As with all weather reports and forecasts, we need additional information to determine the extent and nature of the weather depicted in Fig. 5-13. One of the best sources for the occurrence of precipitation and convection is the Radar Summary Chart. The Radar Summary Chart for this date and time frame is in Fig. 5-14. The strongest activity is along the cold front that extends from Arkansas, through western Louisiana, and into the Texas Gulf coast. In fact there are several solid (SLD) lines of thunderstorms. Convective activity (TRW) extends north into Arkansas. Precipitation wraps around the low-pressure area through Missouri, Kansas, and into western Texas. However, with few exceptions, this activity is light to moderate rain (R) and rain showers (RW), indicating a more stable air mass in these areas.

The cloud mass as seen on the satellite imagery extends to the Rocky Mountains of New Mexico and Colorado. However, the precipitation stops in Kansas and extreme eastern New Mexico. Also note that there are clouds in the upper midwest. These are low clouds, as indicated by the gray color on the IR image, with no precipitation (NE—no echoes) indicated on the Radar Summary Chart.

The proper interpretation of visible and infrared satellite images can aid in the identification of thunderstorms and thunderstorm hazards. Circular or elliptically shaped clouds that appear very cold on IR images may be thunderstorms, especially if they appear bright with irregular texture on visible images. Storms that grow rapidly indicate strong updrafts; these storms are likely to be severe. Storms that become sheared by strong winds aloft in the flight levels can produce severe clear air turbulence at a considerable distance from the storm.

Squall lines are often seen in the cloud patterns ahead of a cold front. In some cases, a thin line of cumulus and cumulonimbus appears ahead of, and parallel to, a frontal band. In other instances groups and clusters of clouds form bands that may have a scalloped appearance.

[Figures 5-15, 5-21, and 5-22 are courtesy of Gary Ellrod of the National Oceanic and Atmospheric Administration (NOAA), National Environmental Satellite, Data, and Information Service (NESDIS).]

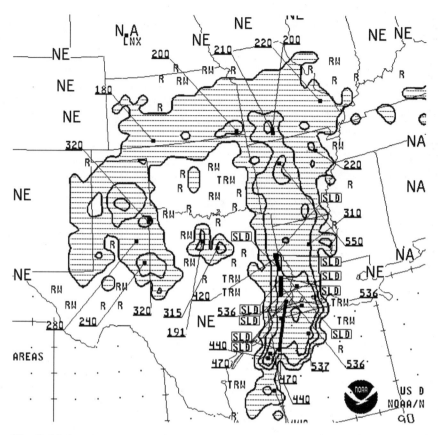

Fig. 5-14 One of the best sources to confirm or refute the occurrence of precipitation and convection is the Radar Summary Chart.

Figure 5-15 shows an early afternoon visible image at 1815Z of a squall line nearing the Mississippi River from western Kentucky to south of Memphis, Tennessee. Storm movement was from the northwest. A gust front has developed and extends from the southern end of the squall line into southern Arkansas. Ahead of the system is an area of moist unstable air, represented by the extensive cumulus clouds. A more stable air mass exists over southwest Louisiana. By 2115Z, the satellite image shows that new convective storms have developed over southern Mississippi and Alabama. These storms are relatively isolated, easily identifiable, and should be circumnavigable. The squall line to the north has strengthened, because of afternoon heating and a sufficient moist, unstable air mass. These storms most likely will not be circumnavigable.

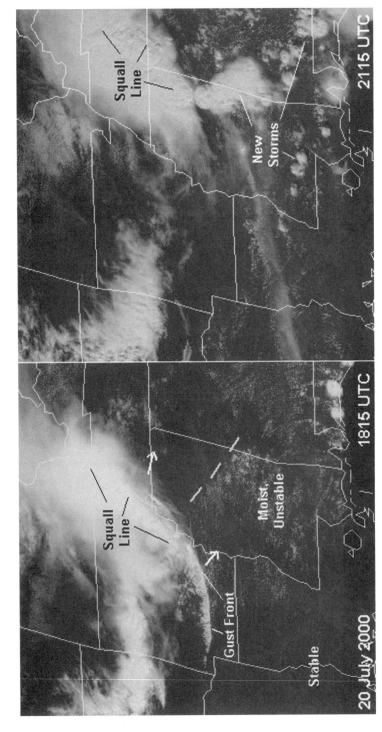

Fig. 5-15 Isolated storms are easily identifiable, and should be circumnavigable; the squall line to the north, however, most likely will not be circumnavigable even with storm avoidance equipment.

Figure 5-16 is a late-summer-afternoon visible image. As is typical this time of year, air mass thunderstorms have developed over the Great Basin and the mountains of central and southern California. There are no overshooting tops, indicating that storms are not severe. Typically winds at mid- and upper levels are light. This is shown by no distinct cirrus blow-off, indicating little upper wind flow and little, if any, thunderstorm movement. Usually these storms are circumnavigable for both VFR and Instrument Flight Rule (IFR) operations. Notice that cirrus tops cover the storms along the southern Sierra Nevada and central California coastal mountains. Without access to real-time weather radar or visual contact with the cells—the ability to maintain visual separation—no pilot should attempt to penetrate these areas either VFR or IFR. These rules apply to any area of air mass thunderstorms.

Fig. 5-16 Air mass thunderstorms are usually not severe and typically can be circumnavigated by both VFR and IFR operations.

Another term used with satellite imagery to identify severe convective activity is the enhanced V. On enhanced IR imagery, a severe thunderstorm often exhibits a V-shaped notch in the top of the cloud. The narrow end of the V points upwind. This signature frequency is usually, but not always, associated with tornadoes, hail, strong surface winds, and extreme turbulence.

Obstructions to Visibility and Fog

In Fig. 5-16, the light gray areas of the Sacramento Valley, northern California, and southern Oregon are smoke caused by the devastating fires of September 1987. Dust storms and sand storms result from strong winds driving particles of dust and sand into the atmosphere. On satellite imagery they are characterized by a dull, hazy, filmy appearance similar to that of thin cirrostratus. The dust or sand can extend over adjacent bodies of water, obscuring the coastline. The edges of dust and sand storms usually taper off uniformly. Dust often shows up better on IR than on visible imagery; and like haze and smoke, dust is best seen during early morning or late afternoon. How can we determine if a particular image is showing haze, smoke, or dust? By comparing the satellite image with current METARs, along with the Surface Analysis and Weather Depiction Charts, the exact phenomenon can be identified.

A flat texture, frequently sharp edges, and patterns conforming to topographical features are the distinguishing characteristics of fog and stratus. It is impossible to distinguish fog from stratus on satellite imagery. When patches of higher clouds are present over stratus and fog, they cast shadows that give the stratus layer a nonuniform appearance. The shadows are usually large and distinct because of the large separation between the low stratus tops and the higher clouds.

Advection, radiation, upslope, and steam fog are usually well defined on visible imagery, except when there is a high cloud cover. In Fig. 5-16, an afternoon visible image, advection fog has cleared most of the land areas, but remains along central California's coastline. Since rain-induced fog is caused by precipitation from a higher layer it does not show up on satellite imagery. Fog dissipates from the edges inward.

Figure 5-17 is a 1-km-resolution visible image for July 10, 2001. Coastal advection fog is typical during the late spring and summer months along the west coast of the United States. Note the detail available with the 1-km image resolution. Depending on pressure gradients that result in onshore or offshore flows, the stratus moves inland during the late afternoon and evening. It is most extensive around sunrise. The coastal stratus usually dissipates from the interior land areas to the coast with daytime heating. Depending on the depth of the layer and pressure gradients, it may clear land areas by midday or remain in the adjacent coastal valley throughout the day. The visible satellite image is an excellent tool to track its progress.

Radiation fog is usually clearly discernible on visible and, at times, on IR imagery, depending on the contrast between surface and cloud top temperatures. Radiation fog, a type of air mass fog, is common to many continental areas. Figure 5-18 shows an extensive area of radiation fog in California's Central Valley. Note the bright, flat appearance, with sharp edges. The edges follow the contour of the

Fig. 5-17 Coastal advection fog is typical in the late spring and summer months along the west coast of the United States. Note the detail available with the 1-km image resolution.

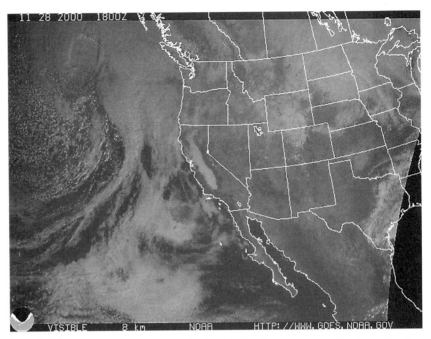

Fig. 5-18 On visible imagery, radiation fog has a bright, flat appearance, with sharp edges that follow the contour of the terrain.

valley. Barely discernible in the picture is what appears to be a small hole in the fog at its northern end. This turns out to be Sutter Butte, elevation 2140 ft. This geographical formation often pokes through the fog.

Radiation fog tends to be patchy and shallow, usually burning off by midmorning. It tends to form in valleys after moisture has been added at the surface from passing storms. As high pressure—clear, stable conditions—builds in an area, circumstances are right for the formation of radiation fog. This condition can become persistent in California's Central Valley during winter and early spring. (In this area the low fog, with tops usually less than 3000 ft, is known as *tule fog. Tule* (tôô lê) is a Spanish word for bullrushes, a marsh plant that grows during this season.) Figure 5-18 shows tule fog in California's Central Valley. Note how well the lateral extent of the fog shows up on satellite imagery. The satellite clearly indicates that the fog extends from the Central Valley through the Carquinez Straits into San Francisco Bay.

A combination of advection, radiation, and upslope cooling produces extensive fog and stratus layers over the central United States. Warm, moist surface conditions exist over the Gulf of Mexico. When this warm, moist air is advected northward, cooled, and lifted, fog and stratus form. The resulting clouds are characterized by a flat texture and sharp edges. When the clouds lay under cirrus associated with the jet stream, shadows from the cirrus give the areas an uneven appearance. The sharp edge of the clouds can often be seen through the jet stream cirrus.

Upslope fog can be widespread and will persist as long as favorable conditions continue. During upslope fog conditions, the VFR pilot is pretty much out of luck. The IFR pilot might not be much better off. He or she might encounter IFR landing minimums, but the condition often exists over areas the size of several states. A legal IFR alternate might be beyond the aircraft's range. Satellite imagery is often useful in determining the extent of upslope fog, as long as higher clouds are not present.

Figure 5-19 illustrates an upslope condition in western Texas and eastern New Mexico. Moist air from the Gulf of Mexico has been advected by the wind toward the Rocky Mountains. Because the terrain rises continuously from sea level at the Gulf coast to around 7000 ft as it approaches the mountains, the air cools adiabatically, and, with sufficient moisture, fog forms. Note how the satellite image depicts the area covered and the extent of this phenomenon.

Turbulence

The presence of turbulence can often be assessed from distinctive cloud forms that appear on satellite imagery. Lenticular cloud formations and wave cloud patterns reveal areas of orographically induced turbulence. Jet stream cirrus locates regions of turbulence induced by strong horizontal shear in the upper troposphere. Convective cloud formations indicate the strength and relative depth of turbulent convection. Turbulence also occurs in certain areas of fronts, vortexes, and vorticity maximums. Satellite imagery can be used to precisely locate these features and the general regions of probable turbulence.

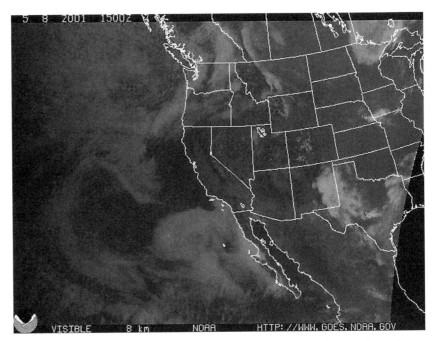

Fig. 5-19 Satellite imagery is often useful in determining the extent of upslope fog, as long as higher clouds are not present.

Figure 5-20 illustrates a mountain wave as seen on visible imagery. Wave activity has developed over the northern Sierra Nevada Mountains and mountains of northeast California and southern Oregon. The wave extends hundreds of miles downstream, as shown in Fig. 5-20, with wave clouds throughout northern Nevada and into northern Utah. The most significant turbulence is close to the mountains, with lesser turbulence farther downstream. Could wave activity be occurring south of Lake Tahoe, east of the southern Sierra Nevada Mountains? Probably. Without sufficient moisture clouds would not develop, but the wave activity may still be present.

Note the many lakes that are clearly discernible on satellite imagery. Figure 5-20 clearly shows the Salton Sea in the southern California deserts, Mono Lake and Lake Tahoe, Walker and Pyramid Lakes in Nevada, and the Great Salt Lake. Figure 5-20 has a higher resolution than that available from the national NOAA Web site. In this midday photo, terrain features are clearly visible. Also, note that many smaller lakes are visible. Dendritic snow patterns of the southern Sierra Nevada

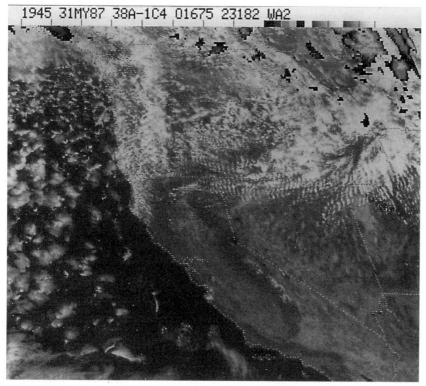

Fig. 5-20 Mountain wave activity is often clearly visible on satellite imagery.

Mountains are easily identifiable. It would seem to be a good flying day, except for the turbulence. All too often this is the case; clear skies and unlimited visibilities mean strong winds and lots of turbulence. Mother Nature never gives us anything for nothing. This also illustrates that the satellite image, in itself, cannot be substituted for a standard weather briefing. It is merely one piece of the "complete picture."

Mountain wave activity also can be seen in Fig. 5-17. Note the wave clouds in the upper right portion of the image. Although significant turbulence is indicated by these clouds, it may not always be present over the entire area affected by the phenomena.

Convective clouds seen on satellite imagery identify turbulent areas. Cumulus and towering cumulus clouds have a typical lumpy, uneven texture and indicate the presence of light to moderate turbulence. Cumulonimbus clouds appearing either as individual cells or in lines or clusters indicate the presence of moderate or severe turbulence. The turbulence associated with cumulonimbus clouds, as we have seen, is not restricted to the clouds themselves, but also occurs many miles from cells in the clear air surrounding the clouds. The specific intensity of turbulence encountered by an aircraft around convective clouds will vary depending on the exact location of the aircraft relative to the cloud, and on the state of development and severity of the cells.

Turbulence is also associated with fronts, developing frontal waves, and areas of positive vorticity advection (PVA). These features have distinctive, recognizable cloud forms that make possible the identification of turbulent areas.

The entire frontal cloud band should be considered as an area with a high probability of turbulence. This is especially true in areas of embedded convective activity. A developing frontal wave and its associated jet stream contain areas of strong horizontal temperature gradient and strong vertical wind shear. Frequently there is a large area of significant turbulence extending through a thick layer of the atmosphere. In this synoptic situation, the turbulence below 20,000 ft is usually associated with frontal clouds and the higher level turbulence related to the jet stream.

PVA maximums are seen on satellite imagery as an area of enhanced cumulus or comma-shaped cloud patterns in the cold air behind the polar front (Fig. 5-1). (Enhanced cumulus on satellite imagery refers to towering cumulus—that is, cumulus with vertical development without cirroform tops, seen on satellite imagery with texture and shadowing.) The favored area for turbulence is the northeast quadrant of PVA maximum. Vortexes that develop within the cold air behind a major cloud band first appear as an area of enhanced cumulus. The comma-shaped cloud indicates the developing stage. As these storms reach maturity, they develop a regular spiral vortex pattern with a long, narrow cloud band similar in appearance to a frontal band.

Large thunderstorms can produce severe to extreme clear air turbulence. Turbulence encountered in clear air, usually above 15,000 ft and associated with wind shear, is classified as clear air turbulence (CAT). Figure 5-21 shows a large severe storm in western Kansas. An overshooting cloud turret—within the circle—identifies the most intense part of the storm. Small-scale wave clouds, known as *billow clouds,* are observed near the cloud top. This is most likely an area of significant turbulence. On the north side of the system, cirrus cloud bands oriented perpendicular to the upper flow indicate another area of significant turbulence. A Boeing 767 reported moderate to severe turbulence in southern Nebraska in the thin cirrus just north of the main anvil cloud. Since intense storms act as a barrier to the air flow, stronger than forecast winds and wind shear may be found on either side of the visible cloud system.

Figure 5-22 is a visible satellite image where a Boeing 727 encountered extreme turbulence near Alma, Georgia. The incident resulted in

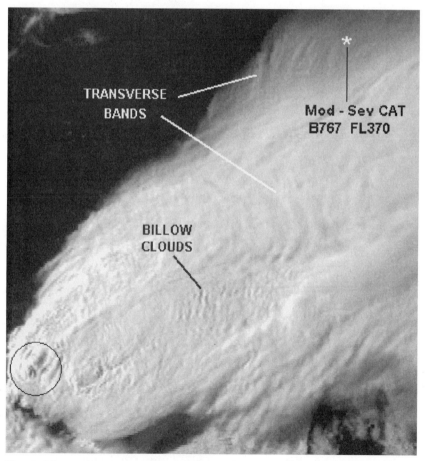

Fig. 5-21 Overshooting cloud turrets, billow clouds, and transverse bands indicate locations of significant turbulence associated with a thunderstorm.

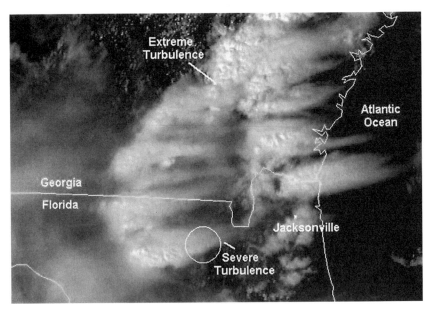

Fig. 5-22 Strong winds aloft tend to increase the severity of thunderstorms and their associated turbulence.

several severe injuries. The incident occurred in an area of thunderstorms with tops to approximately 35,000 ft. Winds near this altitude were about 115 knots. [If we apply our rule of thumb (1000 ft above the thunderstorm cloud tops for each 10 knots of wind), a safe altitude would be 46,500 ft!] The image shows several scattered to broken lines of convective storms, with the cirrus anvils extending well downstream. An additional report of severe turbulence at 33,000 ft was made in northern Florida near one of the cirrus plumes. The circle shows billow wave clouds, indicating possible strong wind shear. Farther north, in the vicinity of Alma, Georgia, there was considerable cirrus from upstream convective clouds, and beneath the cirrus a new line of convection was rapidly forming.

Except where cumulonimbus clouds are present, most turbulence of moderate or greater intensity that occurs above 30,000 ft is associated with the jet stream. Recall that cirrus is often present on the equatorial side of the jet axis. Turbulence occurs on both sides of the sharp edge of the cirrus shield or its associated shadow line and is generally confined to within 180 nm of the jet axis. There is a higher probability of significant turbulence with dense overcast jet stream cirrus than with thin, scattered or broken cirrus. Scattered cirrus seen on satellite imagery is generally associated with smooth to light turbulence. Moderate turbulence is generally encountered in the top two-thirds of a heavy cirrus overcast, associated with the polar jet. The greatest probability of moderate or greater turbulence is associated with the lower two-thirds of the dense overcast cirrus layer of subtropical jet streams. Light turbulence is generally found in the lower third of Polar Jet Stream cirrus and in the upper third of Subtropical Jet Stream cirrus. A higher risk of severe or extreme turbulence exists when transverse bands or cloud trails are observed near the jet.

CHAPTER 6

Satellite Interpretation and Application

As with any weather situation, the key to hazardous weather avoidance is knowledge. This knowledge consists of a thorough understanding of basic weather phenomena, a complete weather briefing just prior to departure, and frequent weather updates en route. The satellite image can be a significant part of this process.

In this chapter we will apply satellite imagery, and the knowledge gained in the two previous chapters, to various aviation weather hazards. For our purposes we will divide hazards into six categories. These are certainly not all the weather hazards with which pilots must contend, but those where satellite imagery can be applied. The categories are

- Low ceilings and low visibility

- Winds and turbulence

- Icing

- Thunderstorms

- Fronts

- Nonfrontal weather systems

From the discussion we'll see that some hazards or potential hazards have direct application to satellite imagery and others have only indirect satellite application.

Low Ceilings and Low Visibilities

Almost one-third of the pilots involved in low-ceiling accidents and two-thirds of those involved in low-visibility accidents had no record of a weather briefing. And, although there are no specific records, it's doubtful they obtained weather en route. Taking everything into account, virtually all low-ceilings and low-visibility accidents are preventable.

> **CASE STUDY**
> The pilot departed a southern California airport during the evening for a flight to Monterey. Arriving in the Monterey area about midnight, the pilot found the airport overcast. The pilot landed on a highway south of the Reid-Hillview airport in the San Francisco Bay area. There was no record of the pilot checking en route for a weather update, which would have revealed the onset of coastal stratus! There is no rational reason for this accident to have occurred. The pilot landed safely and was applauded by some. Unfortunately for aviation, opponents of Reid-Hillview cited it as another reason to close the airport.

In the preceding example, even the required fuel reserve may not have been adequate. A 45-minute fuel reserve doesn't make any sense with the nearest suitable alternate 50 minutes away. What would have happened if the coastal stratus extended beyond the airplane's range? Most likely the accident would have been fatal! How could this incident have been prevented? Simple; update weather en route with Flight Watch or Flight Service. Even though the flight was at night, IR satellite imagery can still often be used in non-daylight hours to determine the extent of the coastal advection stratus.

When conditions are favorable for coastal stratus, VFR pilots should plan arrivals and departures during the afternoon hours. If this is not possible, moving the aircraft to an airport a few miles inland will often allow a morning departure or evening arrival.

> **CASE STUDY**
>
> It's often hazy in the southern portion of California's San Joaquin Valley. A pilot flew for over an hour looking for the Porterville airport. The pilot ran out of fuel and landed in a field. In spite of talking to UNICOM, this pilot could not locate the airport. A week or so later another pilot had a similar experience, under the same weather conditions. Unable to locate the Porterville airport, this pilot called Air Traffic Control. Specialists at the Flight Service Station provided a DF (Direction Finder) steer to Bakersfield.
>
> The FAA has personnel and equipment ready and waiting. But, the pilot does have to ask for assistance. Let ATC help before an incident becomes an accident.

Every year, pilots become lost, even lose control of the aircraft, flying in reduced visibilities. Often flight conditions improve by climbing to a higher altitude. Once above the layer, slant range visibility is usually greater with a distinct horizon preventing disorientation. The seemingly obvious assumption, the closer to the ground the better to see it, usually isn't true. Satellite imagery can often be used to determine the extent of haze and smoke layers.

> **CASE STUDY**
>
> This pilot was operating in the southeast United States, in an area of reduced visibility in haze and fog. The non-instrument-rated pilot had difficulty maintaining heading and altitude. Approaching the airport to land, the aircraft was seen in a descending right turn. The pilot crashed into trees and then hit the ground. The crash was fatal. Visibility in the vicinity of the crash site was estimated to be between 1 and 2 miles in fog. The pilot told several others of concern about flying in hazy conditions.

This pilot's first error was flying in conditions below legal weather minimums. The pilot was obviously uncomfortable about the flight. If that's the case, why depart in the first place? What can be so important as to warrant endangering your life? If any doubt exists, don't go! Second, the pilot could have climbed to a higher altitude, above the haze and fog layer. This may have allowed the pilot to maintain aircraft control. Once on top, the pilot could have obtained the services of ATC to locate or assist in finding a suitable landing site.

Figure 6-1 shows an area of fog that has developed in southeast Texas. The following are weather advisories issued by the Aviation Weather Center (AWC)—AIRMET SIERRA—and the Houston Center Weather Service Unit (CWSU)—ZHU CWA 201.

```
DFWS WA 121345
AIRMET SIERRA UPDT 3 FOR IFR VALID UNTIL 122000


NO SGFNT IFR EXP.
....=
ZHU CWA 201 VALID UNTIL 121445
FROM LFK-50SE IAH-70SW IAH-80NW IAH-LFK
AREA VSBYS 1-3SM BR...LCL CIGS AOB 005. CONDS
IMPRVG BTWN 1430-1530Z.
LP
```

The AWC indicates that no significant IFR conditions are expected. However, from the satellite image in Fig. 6-1 we see that indeed radiation fog has developed. This is reflected in the Houston CWA, which advertises areas of visibilities 1 to 3 miles in mist, with local ceilings at or below 500 ft, and with the conditions improving between 1430Z and 1530Z. This is not inconsistent, but reflects the differing purpose and scope of different aviation weather products. The AWC will not amend AIRMET SIERRA because the phenomenon is not expected to last. But the CWSU has issued an advisory because it is in its area of jurisdiction and the CWA scope allows the issuance advisories for localized areas and short time frames. This is also an excellent example of satellite imagery's use in aviation weather forecasts.

CASE STUDY

With low ceilings and visibilities, the greatest hazard occurs when both prevail. We were on a flight from Shreveport, Louisiana, to Mineral Wells, Texas. Ceilings were below 1000 ft, but visibility was unlimited. In the sparsely populated Texas

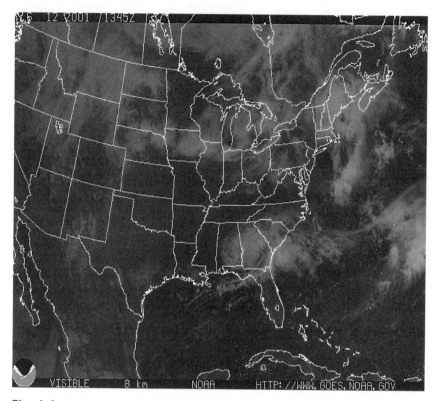

Fig. 6-1. Even though a weather advisory has not been issued, the satellite picture often can reveal an area of potentially hazardous weather.

countryside, it was easy to maintain legal distance from objects on the ground and avoid towers and power lines. As we flew south of Dallas, the clouds lowered to the ground. In the non-instrument-equipped Cessna 150, there was only one course of action—reverse course and land. We were delayed another day before we were able to proceed west.

Figure 6-2 shows a high-resolution (1-km) image of southern California. In the image, it's easy to see just exactly how far the advection stratus has moved inland. It also can be seen that some of the coastal hills are above the stratus, which confirms the relatively low stratus tops. With the use of an aeronautical chart, pilots can pinpoint those airports in the clear and those affected by the fog.

Fig. 6-2. With a high-resolution satellite image and an aeronautical chart, pilots can pinpoint those airports in the clear and those affected by the fog.

Even pilots with limited instrument capability—I'm speaking of aircraft and equipment—have the option of filing IFR. Always the best situation is to "get into the system" before reaching IFR conditions.

> **CASE STUDY**
> When I first came into the FAA I was assigned the Lovelock, Nevada, FSS. I would routinely fly from Lovelock to Van Nuys, California. My Cessna 150, although equipped and certified for IFR, did not have the capability to fly IFR over the Sierra Nevada Mountains. But, if the only weather was coastal stratus, I would file VFR to Palmdale and IFR from Palmdale to Van Nuys. It's always easier to pick up a prefiled clearance before reaching congested terminal airspace.

On one of our trips to Oshkosh, the leg from Mason City, Iowa, to Madison, Wisconsin, was plagued with marginal VFR weather toward the destination. I elected, rather than to try to fly under the weather or attempt to pick up IFR in the Madison area, to file IFR from Mason

City. The weather was clear, so we departed VFR and picked up the clearance with Center during climbout. It was a good plan. With marginal weather in the Madison area controllers were refusing pop-ups and instructing VFR aircraft to remain clear of Class C airspace. Pilots were advised to contact the next sector for traffic advisories. This is not a criticism of the controllers. There is only so much airspace, especially in terminal areas.

> ## CASE STUDY
>
> "Plan A" doesn't always work. After remaining overnight at North Platte, Nebraska, we planned to continue westbound to Cheyenne. A front had passed through the previous day, and the ground was moist. Where is the airport? Of course, next to the river where the fog and low clouds are usually the worst. The satellite image indicated that stratus was restricted to the river valley and about 15 miles to the west. Beyond that point the weather to the west was good and I obtained a special VFR clearance. After takeoff it became apparent it wasn't going to work. We had no option except to land. We returned to the airport office, filed an IFR flight plan, and 20 minutes later were on our way. (When you have time to spare, go by air.)

An AIRMET was in effect for mountain obscurement in Idaho. During the briefing for the flight from Winnemucca, Nevada, to Idaho Falls, the briefer advised that VFR flight was not recommended (VNR). This was a little silly because our flight took us through the Snake River Valley, not the mountainous areas that were obscured. On opening our flight plan and, again, updating weather en route we were VNR'd. The valley was perfectly fine. The satellite image confirmed that the mountains were indeed obscured and that the valley was clear.

The next morning we prepared to fly from Idaho Falls to Billings, Montana. Weather at both ends was good, with no weather advisories for the route. As we approached Yellowstone National Park, the valleys were fogged in, and there were several layers aloft. Most of the mountains were obscured. Because of the minimum en route altitudes and freezing levels, IFR (on purpose) was not an option. We picked our way through, at times climbing to over 13,000 ft in the Cessna 172. We always had the option to reverse course and return to Idaho Falls or divert into West Yellowstone. We were lucky and on the north side of the park were able to descend below the clouds, verify with Flight Watch that Billings weather was good, and proceed to our destination. This was a case where the satellite image was of no value in determining valley fog; there was a higher cloud layer obscuring the lower clouds.

This scenario has the potential for disaster. Navigation was difficult because of sparse navigational aids and terrain obscurement. As is my practice, I had a dead reckoning course planned and calculated. Throughout the flight I was evaluating options on the basis of weather and fuel: proceed, return, or divert.

Oh, a side note. We were not VNR'd on the flight to Billings. There were no weather advisories. The point is that the existence, or lack, of an advisory does not preclude, or guarantee, a safe flight. Pilots must evaluate each flight separately, on the basis of the reported and forecast weather, and their aircraft, training, and experience.

Recall that we mention that no segment of aviation is immune from low-ceiling and -visibility accidents. The typical IFR accident scenario is descending below minimums. It has happened to the airlines and military, as well as general aviation pilots. For example, a few years ago a United States Air Force T-43 crashed in Bosnia. Investigators discovered, among other things, that the aircraft was not properly equipped for the approach to be used and not flown at proper airspeeds. It's foolhardy to attempt an approach without the appropriate equipment functioning normally. This includes appropriate and current charts. During the approach, if anything isn't normal abandon the approach!

The dos and don'ts of flying in low ceilings and visibilities:

- Do obtain a complete weather briefing.

- Do update weather en route.

- Do get into the IFR system before you encounter poor weather.

- Do ask for help whenever the situation becomes doubtful or uncertain.

- Do have options and a plan for each option.

- Don't let the briefer make the decision (VNR or no VNR).

- Don't fly below legal or personal minimums.

- Don't run out of options; land before you do.

Winds and Turbulence

Knowledge is the key to avoiding strong or gusty surface wind accidents. Three areas of knowledge are required. First, know your own limitations. Second, know the performance and limitations of your aircraft. Third, know the surface wind conditions. Surface winds are not directly observed on satellite imagery. However, by viewing a loop, cloud movements can be used to obtain an indirect indication of wind direction and speed. A knowledge of cloud tops is also required. In fact, the National Weather Service uses satellite-derived wind on a number of its products, specifically constant-pressure charts over the oceans.

Typically, aviation accidents are not directly related to turbulence. However, turbulence is a contributing factor to spatial disorientation, loss of control, and structural damage. Therefore, a sound understanding of turbulence, where it is likely to occur, and strategies to reduce its effects are essential.

The intensity of turbulence is, to some degree, affected by aircraft type and flight configuration. United States Air Force studies have shown the following to generally increase the effects of turbulence.

- Decreased weight

- Decreased air density

- Decreased wing sweep angle

- Increased wing area

- Increased airspeed

Classifications for the intensity of turbulence can be found in the *Aeronautical Information Manual* and *Aviation Weather Services*.

An object placed in any moving air current impedes the flow, causing abrupt changes in wind direction. As the current closes in behind the object, eddy currents develop leeward of the obstruction. This turbulence is caused by the obstruction and not by any meteorological phenomena inherent in the air itself. Turbulence caused in this manner is termed *mechanical.*

Air flowing through mountainous terrain is forced upward on the windward side and spills downward over the leeward side. The degree of turbulence induced by the mountains depends on the shape and size of the mountains, the direction and speed of the wind, and the stability of the air. Downdrafts on the leeward side may be dangerous and can place an aircraft in an attitude from which it may not be able to recover.

When winds aloft blow in excess of about 40 knots, approximately perpendicular to a mountain range, and speed increases with height in a stable atmosphere, a mountain wave or standing wave can develop. Turbulence can become severe to extreme. Updrafts and downdrafts occasionally reach 3000 ft/min and can exist to the tropopause or slightly higher. Downdrafts may dip to the surface on the leeward side of the mountains. Large waves may form to the lee of mountains and may extend hundreds of miles downstream. Complete overturning may occur under the wave crests at lower levels.

Major mountain waves occur east of the Cascade Mountains in Washington and Oregon, the Sierra Nevada Mountains in California and Nevada, the Rocky Mountains, and the Appalachian Mountains. Smaller waves can develop over any hill or mountain.

These waves are termed *standing* because the crests and troughs remain stationary while the wind undulates rapidly through them. With sufficient moisture, clouds form. And, although the clouds appear to be stationary, condensation occurs at the leading edge of the cloud and evaporation occurs at the trailing edge. The clouds are continuously forming and dissipating.

Cap clouds hug the tops of the mountains and appear to flow down the lee side. They are indicators of strong downdrafts on the lee side of the mountain range.

Lenticular clouds are a visual identification of the wave crests. They may extend to 40,000 ft. Depending on the vertical distribution of moisture in the atmosphere, these clouds may be stacked. The distance from the ridge to the first wave crest depends on wind speed, lapse rate, and the shape of the ridge.

Rotor, bell-shaped, clouds that often appear as tubular lines of cumulus or fractocumulus clouds parallel to the ridge line underneath the lenticulars, always imply severe or greater turbulence. Depending on the strength of the flow and atmospheric conditions, less-developed lines may form downstream beneath the crests of subsequent waves. The base of the rotors is about the same height as the ridge and may extend vertically 3000 to 5000 ft. Violent updrafts are common in the vicinity of rotor clouds.

Figure 6-3 shows visible and infrared images of mountain wave activity in northern Nevada. The wave pattern can be seen on the visible image; since the IR image indicates tops, relatively cold, these are most likely midlevel clouds—standing lenticular altocumulus (ACSL).

Below are excerpts from the Salt Lake City Area Forecast and AIRMET Bulletin and the Winds and Temperatures Aloft forecasts for the day and time frame of the satellite images in Fig. 6-3.

```
SLCC FA 121045
SYNOPSIS AND VFR CLDS/WX
SYNOPSIS VALID UNTIL 130500
CLDS/WX VALID UNTIL 122300...OTLK VALID 122300-130500
ID MT WY NV UT CO AZ NM


SYNOPSIS... MOD CYC WLY FLOW WL CONT S OF THE UPR SYS
ACRS CNTRL/SRN SXNS OF THE FA AREA. SFC...WK LOW IS OVR
CNTRL WY WITH CDFNT THRU CNTRL UT-SRN NV.

SLCT WA 121345
AIRMET TANGO UPDT 2 FOR TURB VALID UNTIL 122000


AIRMET TURB...ID WY NV UT CO AZ NM WA OR CA
FROM YDC TO DNJ TO BOY TO CYS TO TBE TO INK TO 50S TUS
TO MZB TO 30W RZS TO PYE TO FOT TO BLI TO YDC
```

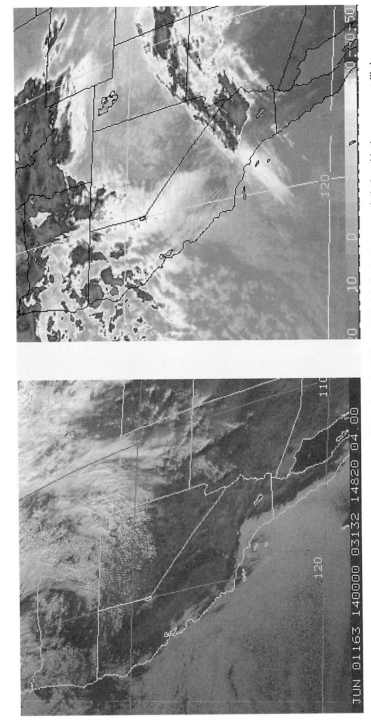

Fig. 6-3. The wave pattern can be seen on the visible image (left); since the IR image (right) indicates tops are relatively cold, these are most likely midlevel clouds—standing lenticular altocumulus (ACSL).

OCNL MOD TURB BLW 160 EXC BLW FL180 OVR CO AND NRN
NM. CONTG BYD 20Z THRU 02Z.

.

DATA BASED ON 120000Z
VALID 121200Z FOR USE 0900-1800Z. TEMPS NEG ABV 24000

FT	6000	9000	12000	18000	24000	30000
RNO	2912	2828+04	3035+00	2956-13	2976-23	298737
BAM		2920+01	2733-03	2749-16	2868-27	278840
ELY		3019+05	2730+01	2546-12	2656-24	277139

The synopsis indicates that a moderate cyclonic westerly flow will continue south of the upper system across central and southern sections of the FA area. At the surface a weak low is over central Wyoming with a cold front through central Utah and southern Nevada. Note that the system is so weak in southern Nevada that there are no clouds associated with the front.

AIRMET Tango calls for occasional moderate turbulence below 16,000 ft for Nevada. This is supported by the Winds Aloft forecasts. The winds are averaging westerly at 25 to 50 knots, between 12,000 and 18,000 ft. This is all consistent with the satellite imagery, terrain, synopsis, and forecasts.

Daytime heating causes rising air currents that produce thermal turbulence—also called *convective turbulence.* Thermal turbulence usually occurs within 7000 ft of the surface in stable or conditionally unstable air. This means that vertical movement requires an initiating force, in this case surface heating. In stable or extremely dry air, skies remain clear. Should air parcels reach the lifted condensation level, saturation occurs, and stratocumulus, or fair weather cumulus, clouds will form. These clouds are most often scattered, and rarely become overcast. There are rising air currents in the clouds and descending currents in the clear air. Flight will be turbulent below the clouds and smooth above the clouds. The air is stable, since there is little vertical development. This would typically manifest itself as an open cellular pattern on satellite imagery.

Should the air be conditionally unstable—a parcel of air that becomes unstable on the condition it is lifted to the level of free convection (LFC)—cumuliform clouds form, which can develop into air mass thunderstorms. Instead of the scattered stratocumulus or fair weather cumulus, towering cumulus will develop; often accompanied by rain showers and thunderstorms.

Although thermal turbulence rarely becomes severe, it can be extremely uncomfortable and annoying. Because thermal turbulence is caused by surface heating, it can usually be avoided by flying before midmorning or waiting until late afternoon. Otherwise, the only remedy is to climb above the turbulent layer, which might be marked by clouds. A word of caution to the VFR pilot: If you elect to fly above the clouds, be careful not to get caught on top. Should the air mass be conditionally unstable, clouds can build at an alarming rate and close up even faster.

Wind shear–induced turbulence is caused by frontal systems, and is associated with clear air turbulence and thunderstorms. It is also associated with inversions and evaporative cooling.

Inversion-induced wind shear turbulence develops along the boundary between cool air trapped near the surface and warm air aloft. The turbulence tends to be strongest in valleys during morning hours when temperature differences are greatest. Moderate or greater turbulence may be encountered when you are penetrating the layer. After an initial outside air temperature rise during the climb through the haze, the air on top will be clear and cool. It is possible to have several haze or smoke layers trapped in inversions aloft.

Another type of wind shear, which occurs in a vertical plane rather than a horizontal one as inversion-induced shear does, is evaporative cooling turbulence. Evaporative cooling turbulence develops in areas of precipitation. This usually occurs in a dry environment with convective activity. Precipitation evaporates and cools the air, causing downdrafts. A pilot penetrating these areas will encounter wind shear turbulence, which can be severe. Turbulence can be avoided by circumnavigating areas of precipitation.

We will discuss the implications of thunderstorms and fronts on satellite imagery in subsequent sections. Like mechanical turbulence, inversion-induced turbulence is not reflected in satellite imagery.

Evaporative cooling manifests itself as clouds, typically by towering cumulus on both visible and IR satellite images. Recall from Chap. 5 that probable areas of jet stream turbulence may show up on satellite images.

Icing

Structural icing accidents accounted for only about 40 percent of total accidents involving icing. A majority of icing accidents are attributed to carburetor or induction system icing, with less than 10 percent involving icy runways. Most structural icing accidents occurred when the pilot continued flight into known icing, severe weather, or deteriorating weather conditions. A lesser amount occurred on approach or landing in icing conditions or with ice accumulation.

The object with icing is to minimize exposure. At temperatures of 0°C or less, avoid flying in clouds or precipitation. Should ice be encountered, immediately notify ATC and initiate a plan of action. The first consideration, if the aircraft has sufficient performance, might be to climb to colder air or above the clouds as indicated by PIREPs and the weather briefing. Ice will slowly sublimate—change from a solid directly into a gas—when on top. Or, descend to warmer air as indicated by the actual freezing level on climb out. You did note the freezing level on climb out? This requires a careful check of minimum en route altitudes (MEAs). Finally, if you have to, turn around; presumably you came from an ice-free area. The point is, do something!

CASE STUDY

A nonturbocharged Baron, without ice protection equipment, departed Reno, Nevada, for southern California. Moderate icing and severe turbulence were forecast. The pilot elected to fly a direct course along the crest of the Sierra Nevada Mountains, the route where the most intense icing and turbulence could be expected. The aircraft iced up, resulting in a fatal accident.

The pilot had no way out because the MEA was the aircraft service ceiling. The terrain was well above the freezing level and the pilot failed to reverse course at the first sign of ice. What other options were available? The pilot could have crossed the mountains near Sacramento, minimizing exposure to ice, and once over the Sierras, it was all downhill. The pilot could have flown toward Las Vegas where the weather was considerably better, or simply waited for better weather conditions. Certainly access to a current visible satellite image would have graphically shown the areas of greatest hazard and the most expeditious route to clear skies.

When the Baron became ice covered, the pilot had no option but to ride it to the crash site. Attempting flight under these conditions and with this type of equipment was quite literally suicide.

> **CASE STUDY**
> A Bonanza pilot departed the San Francisco Bay area on a flight to Los Angeles. Icing above 7000 feet was forecast and reported. The pilot elected to fly at 11,000 feet. The pilot's last words were, "I've iced up and stalled." The crash occurred in the San Joaquin Valley where the elevation was near sea level. Minimum altitudes in the vicinity of the crash were well below the freezing level. The pilot simply did nothing until aircraft control was lost.

If a descent through an icing layer is required, remember the objective is to minimize exposure. Under such circumstances, negotiate with ATC to obtain a continuous descent. Avoid, if possible, level flight in clouds. ATC is usually very responsive to such requests.

A popular notion in some aviation circles is that a pilot's mere mention of ice will receive emergency-like handling. Icing might be an emergency, but remember the controller's job is to separate aircraft within a finite amount of airspace. ATC might have to assign a higher altitude, but ATC cannot, and should not, be expected to fly the aircraft, or assume the responsibility of pilot in command. To paraphrase: An accurate pilot report of actual icing conditions is worth a thousand forecasts.

In freezing precipitation, aircraft without a heated pitot and alternate static source, especially in IFR conditions, would be in serious trouble. Another significant factor, especially for aircraft without ice protection equipment, is that accumulated ice could be carried all the way to the ground, making landing extremely hazardous. It cannot be overemphasized that this hazard can affect VFR as well as IFR operations. Should this phenomenon be encountered in aircraft without ice protection equipment, virtually the only option, and certainly the safest, is to fly into warmer air and land. Pilots who fly into ice, with aircraft not certified for flight in icing conditions, must in every sense of the words, have the "right stuff." Because they become test pilots!

Continually updating the weather picture is the key to managing a flight, especially in aircraft with limited or no ice protection equipment. Icing can be significant during descent, especially when destination temperature is at or below freezing. Flight Watch can provide information on tops, temperatures aloft, reported and forecast icing, and current surface conditions.

CASE STUDY

BFL UA/ OV EHF/TM 1900/FL100/TP UNKN/SK SKC/IC MOD
Could this be an example of "clear air icing"? No. A strong weather system was forecast to move rapidly into central California. As it happened, the system stalled off the coast, with weather advisories for mountain obscuration and icing continuing in effect. When it became apparent the system had stalled, I called the Aviation Weather Center and asked the forecaster to amend the advisory for mountain obscuration. I didn't mention icing; I assumed it would be amended as well. Silly me! Sure enough the forecaster amended the mountain obscuration, but left the icing advisory in effect. On the basis of the revised TAFs and the satellite picture, I stopped issuing the icing advisory, but one of my coworkers vented some frustration in the form of this PIREP.

The following weather event occurred on December 21, 1998. A pictorial view of the weather is often helpful in developing the "complete picture." For this we need access to graphic products. With FAA and NWS consolidation, a visit to an aviation weather facility is usually not practical. But, with DUATs and the Internet, graphic products are becoming more accessible. Obtaining graphic products prior to the standard Flight Service Station briefing provides a general picture of the weather. Like checking terrain, altitudes, and airport information, a preliminary look at the weather provides a helpful background for the preflight briefing.

Figure 6-4 contains a morning Weather Depiction and Radar Summary Chart for December 21, 1998. An arctic cold front extends southeast of our route from Goodland to Wichita, Kansas. The general weather along the route: departure VFR, en route MVFR, and destination IFR.

Freezing drizzle in southeast Kansas is a red flag for supercooled large drops (SLD)—a severe icing hazard. Station models show freezing drizzle behind the front from the Texas panhandle to Missouri. This location coincides with one of the typical SLD locations: 25 to 130 miles behind an arctic front. The arctic cold front is undercutting and lifting warmer air to the south. This feature is often referred to as *overrunning warm air aloft*. Precipitation falls as liquid into colder air below, then freezes on contact with a surface that is below freezing.

With a knowledge of weather patterns, we would expect improvement from the north during the day—we'll verify this with the Area Forecast (FA) and Terminal Aerodrome Forecasts (TAFs). If we're looking for alternates, north appears to be the best bet. Upslope continues over the plains of Colorado, with snow and IFR conditions—not favorable for an alternate. This is confirmed by the satellite images in Fig. 6-5. (The upper image in Fig. 6-5 is infrared and the lower image is visible.) Widespread areas of freezing drizzle and IFR conditions continue south and east of a route from Goodland to Wichita, Kansas. Even though we may find legal alternates in this area, they might not be the best choice. Legal does not necessarily mean safe.

The radar chart is encouraging. It depicts scattered light to moderate precipitation, in the form of rain or rain showers. Rain indicates a stable air mass—typically less serious icing. The tops of precipitation are in the 6000- to 8000-ft range. With relatively stable air, we would expect cloud tops to be within several thousand feet of radar tops, in the 9000- to 12,000-ft range. What about precipitation in

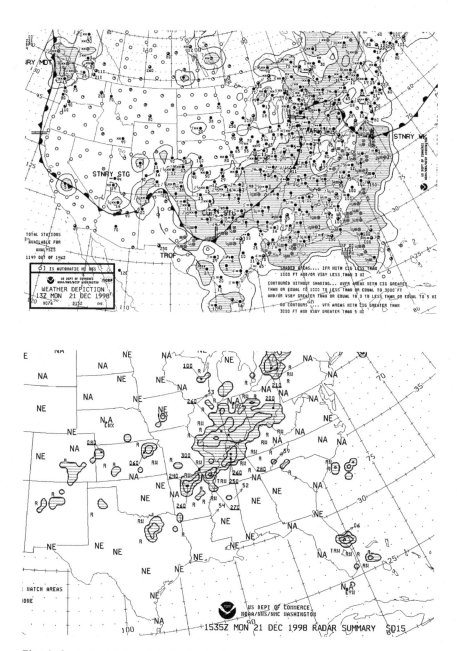

Fig. 6-4. A pictorial view of the weather is often helpful in developing the "complete picture."

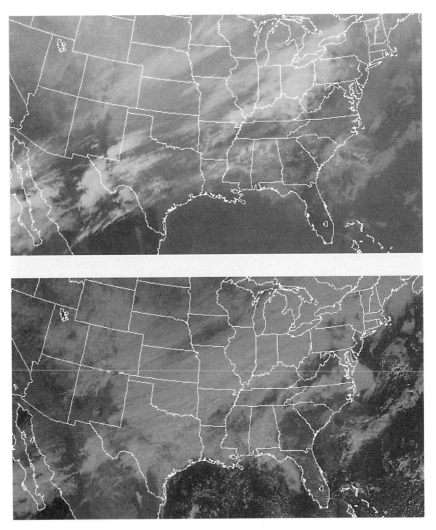

Fig. 6-5. These satellite images (upper infrared, lower visible) indicate widespread clouds, but relatively low tops, with no convective activity along the route from Goodland to Wichita.

the vicinity of Wichita? The radar chart shows the symbol NA. For some reason, the data is Not Available. With no echoes (NE) reported in Oklahoma and the weather depiction showing light, steady precipitation, we could reasonably expect the conditions shown in western Kansas to extend into the Wichita area. Note that from the Weather Depiction and Radar Summary Charts in Fig. 6-4 that the front is most active from northern Arkansas through the Ohio Valley.

Satellite imagery in Fig. 6-5 confirms our interpretation of the Weather Depiction and Radar Summary Charts: widespread clouds, but relatively low tops, with no convective activity along the route. What about the clouds over Nebraska? The weather depiction shows mostly clear conditions. These are most likely automated observations, even though not depicted as such, which report maximum cloud heights of 12,000 ft. The visible image shows relatively thin clouds, the IR image shows cold tops. This is most likely a cirroform layer.

With the preceding background, we're ready for the briefing from either FSS or DUAT. Each will contain essentially the same information. We'll call the FSS or log onto the computer, provide the necessary background information, and request a standard briefing. The DUAT briefing will appear in the following sequence.

The first product displayed is the Area Forecast.

CHIC FA 211045

SYNOPSIS AND VFR CLDS/WX
SYNOPSIS VALID UNTIL 220500
CLDS/WX VALID UNTIL 212300...OTLK VALID 212300-220500
ND SD NE KS MN IA MO WI LM LS MI LH IL IN KY

SYNOPSIS...10Z LOW PRES NRN LM WITH CDFNT TVC-SBN-ARG. HI PRES NERN SD. BY 05Z CDFNT OVR XTRM ERN KY. HI PRES SERN KS. LK EFFECT SHSN/BLSN OVR GRTLKS THRU 05Z.

KS
SERN...CIG OVC010-020 TOP 120. OCNL VIS 6-5SM -FZDZ BR. 15-17Z FZDZ BECMG SN. 21Z AGL SCT-BKN020 CIG OVC040 TOP 080. OTLK...VFR.
WRN...CIG BKN010-020 TOP 120. VIS 6-5SM -SN. 18Z AGL SCT-BKN020 CIG BKN040. OCNL -SN. 22Z AGL SCT-BKN040 BKN100 TOP 150. OTLK...VFR.
CNTRL/NERN...CIG BKN-SCT010-015 OVC030 TOP 100. OCNL VIS 6-5SM -SN. 20Z AGL SCT025 CIG BKN040 TOP 100. OTLK...VFR.

The synopsis discloses a cold front east of our route in southeastern Missouri, moving eastward, high pressure over northeastern South

Dakota. The synopsis confirms our analysis of the Weather Depiction and Radar Summary Charts. The front is not mentioned in Oklahoma and Texas because it is relatively weak and diffuse and areas are not within the Chicago FA coverage. The front is, however, strongest in the Ohio Valley and Arkansas, even producing some thunderstorms. The front is forecast to move eastward during the period. High pressure is expected to move into southeastern Kansas. This confirms our expectation of better weather to the north.

The route forecast indicates, for western Kansas, scattered to broken clouds 2000 ft AGL, ceilings 4000 broken, tops 12,000 MSL, occasional light snow until mid- to late afternoon; central Kansas 1000 to 1500 scattered to broken AGL, 3000 overcast AGL, tops 10,000 MSL, visibility occasionally 3 to 5 miles in light snow; southeastern Kansas 1000 to 2000 AGL, tops 12,000 MSL, visibilities occasionally 3 to 5 miles in moderate freezing drizzle, becoming snow by 17Z. Conditions improving during the afternoon. The forecast confirms our analysis that, in general, we'll want to avoid altitudes near the cloud tops, the 9000- to 12,000-ft range, unless we can remain clear of clouds.

AIRMET ZULU pertains to the second two-thirds of the flight; occasional moderate mixed or rime icing in clouds and precipitation below 18,000 ft, freezing level at the surface. Why no icing in northwest Kansas? The air is too cold. Cloud droplets and precipitation will be frozen. The AIRMET does, however, imply possible SLD (MXD ICGICIP) within its area of coverage.

Flight into Arkansas and the Ohio Valley presents three additional hazards: widespread IFR conditions, high tops, and thunderstorms. With widespread IFR, a suitable alternate may be beyond the range of most small aircraft. Radar shows precipitation tops well above 20,000 ft. Getting on top may not be possible.

Flight to the west, for example, Denver, would have a different set of hazards. The area is under the influence of upslope. Conditions are IFR, with snow, and relatively low tops. This is again confirmed from the satellite image in Fig. 6-5. Aircraft icing is not a serious factor because of the cold temperature. But low ceilings and visibilities and snow-covered runways persist and will continue as long as upslope remains a factor. Whiteout would certainly be a factor in these areas.

VFR flight would certainly be out. IFR flight would have to contend with conditions close to, or below, minimums over a relatively widespread area. However, suitable alternates exist to the northeast and east well within the range of most aircraft.

Thunderstorms

Accident statistics show that a majority of thunderstorm accidents occur to non-instrument-rated low-time private pilots; over half resulted in fatalities. Ironically, most received a preflight weather briefing. Most occurred when pilots initiated IFR flight into adverse weather, attempted VFR flight into deteriorating weather, or attempted to fly in or around thunderstorms. Some occurred when flight was continued into areas of embedded thunderstorms. Others resulted in loss of control due to high, gusty winds or crosswinds.

The violent nature of thunderstorms causes gust fronts, strong updrafts and downdrafts, and wind shear in clear air adjacent to the storm out to 20 miles, with severe storms and squall lines. Precipitation, which is detected by radar, generally occurs in the downdraft, while updrafts remain relatively precipitation free. Clear air or lack of radar echoes does not guarantee a smooth flight in the vicinity of thunderstorms.

> **CASE STUDY**
> A Bonanza pilot approached an area of thunderstorms in California's Central Valley. The pilot received the latest weather radar and satellite information, as well as PIREPs and surface observations from Flight Watch. The pilot safely traversed the area with minimum diversion or delay.

When it comes to thunderstorms, microbursts, and wind shear: A pilot's proper application of many resources—training, experience, visual references, cockpit instruments, weather reports and weather forecasts—make avoidance possible.

The following weather event occurred on May 26, 2001. Below are excerpts from the Miami Area Forecast (FA) and Convective SIGMET 59E.

```
MIAC FA 261745
SYNOPSIS AND VFR CLDS/WX
SYNOPSIS VALID UNTIL 271200
CLDS/WX VALID UNTIL 270600...OTLK VALID 270600-271200
NC SC GA FL AND CSTL WTRS

.

SYNOPSIS...WEAK FNTL BNDRY AT 18Z ALG A ECG-CHS-CRG-
SRQ LN. BTN 00Z AND 06Z FNT WILL MOV OVR THE CSTL WTRS
OF NC SC GA AND NRN FL.

.

NC
NC W OF A 50S LYH-30N FLO LN...SCT OCNLY BKN040. ARND 00Z
SKC.
OTLK...VFR.
RMNDR NC...BKN030-050 BKN090..OVC CI ABV. TOPS FL250. SCT
TSRA MNLY IN LNS. CB TOPS FL400. TS POSS SEV. 01Z-03Z SCT
CI. OTLK...VFR. 09Z-12Z MVFR BR.

.

SC
SC E OF A 30N FLO-CHS LN...BKN030-050 BKN090..OVC CI ABV.
TOPS FL250. SCT TSRA MNLY IN LNS. CB TOPS FL400. TS POSS
SEV. 23Z-01Z SCT CI. OTLK...VFR.
SC S OF A CHS-IRQ LN...SCT040. 19Z-21Z SCT040 SCT-BKN120.
01Z SCT CI. OTLK...VFR.
RMNDR SC...SCT040. 00Z SKC. OTLK...VFR.


MKCE WST 261755
CONVECTIVE SIGMET 59E
VALID UNTIL 1955Z
DE VA NC SC AND SC CSTL WTRS
FROM 20N SBY-40N ORF-40SSE RIC-40E FLO-30ESE CHS
LINE SEV TS 40 NM WIDE MOV FROM 22030KT. TOPS TO FL450.
HAIL TO 2 IN...WIND GUSTS TO 60 KT POSS.
```

The Miami FA synopsis reveals a weak frontal boundary along a line through Elizabeth City, North Carolina; Charleston, South

Carolina; Jacksonville, Florida; and Sarasota, Florida. The front is expected to move over the coastal waters of North Carolina, South Carolina, Georgia, and northern Florida between 00Z and 06Z.

The forecast for North Carolina west of a line 50 nm south of Lynchburg, Virginia, and Florence, South Carolina: 4000 scattered occasionally broken, becoming clear around 00Z. The forecast for the remainder of the state: 3000 to 5000 broken, 9000 broken, overcast cirrus above, with scattered thunderstorms and rain showers mainly in lines, cumulonimbus tops to 40,000 ft, thunderstorms possibly severe, conditions improving between 01Z and 03Z.

The forecast for South Carolina east of a line 30 nm north of Florence to Charleston: 3000 to 5000 broken, 9000 broken, overcast cirrus above, tops to 25,000 ft, scattered thunderstorms and rain showers mainly in lines, cumulonimbus tops to 40,000 ft, thunderstorms possibly severe, improving between 23Z and 01Z. The forecast south of a Charleston to Colliers line and the remainder of South Carolina: no significant weather.

Convective SIGMET 59E covers a line from 20 nm north of Salisbury, Maryland, to 40 nm north of Norfolk, Virginia, to 40 nm south-southeast of Richmond, Virginia, to 40 nm east of Florence, South Carolina, to 30 nm east-southeast of Charleston, South Carolina, 40 nm wide. This is a line of severe thunderstorms moving from 220° at 30 knots, with tops to 45,000 ft; hail to 2 inches and wind gusts to 60 knots are possible.

The line of severe thunderstorms advertised in the Convective SIGMET can clearly be seen on the visible satellite image in Fig. 6-6 and the infrared image in Fig. 6-7. Notice how the Area Forecast correctly predicts this phenomenon, but because of its scope and limitations is unable to more precisely describe the location of the severe weather. However, it is within the scope of the Convective SIGMET to much more precisely define the area affected. The satellite images confirm the more accurate location of the phenomenon and can tell us when the severe weather has passed. Additionally, the satellite images show that there is not a second band of weather behind the front. This implies that, once the weather has passed, we should be able to conduct both routine VFR and IFR operations.

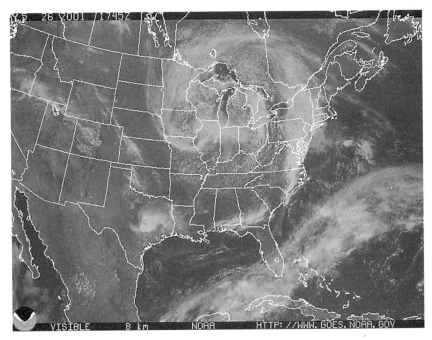

Fig. 6-6. The line of severe thunderstorms advertised in the Convective SIGMET can clearly be seen on this visible satellite image.

CASE STUDY

An intensity 5 thunderstorm was 10 miles west of the airport moving east, toward the airport. Clouds associated with the cell created an overcast layer above the airport. ASOS reported the weather as visibility 10 miles, clear. This was not very representative of the weather at the time.

Many ASOS sites are not yet equipped to report thunderstorms. This is typically indicated in the remarks of the observation as "...RMK TSNO." If the clouds were above 12,000 ft, the ASOS would report "...CLR...." Pilots, dispatchers, and FSS briefers need to understand the limitations of all weather reports and forecasts; automated observations are no exception. As required by regulations, all available information must be analyzed and applied to every flight.

A major criticism of automated observations is their inability to detect freezing precipitation or thunderstorms. Steps are currently

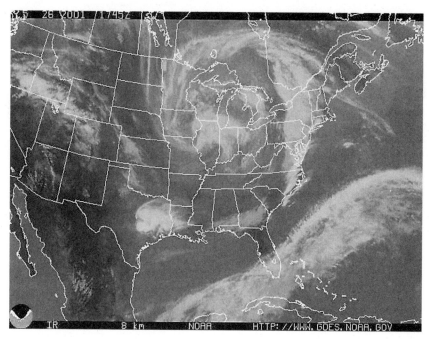

Fig. 6-7. The infrared satellite images confirm the more accurate location of the phenomenon than is available in the Area Forecast, and can tell us when the severe weather has passed.

being taken to alleviate these limitations. Automated sites are being equipped with freezing precipitation sensors. This will allow automated stations to report the occurrence of freezing rain and freezing drizzle. The Automated Lightning Detection and Reporting System (ALDARS), which acquires lightning information from the National Lightning Detection Network, will allow AWOS/ASOS to report the occurrence of a thunderstorm. ALDARS is operational at numerous AWOS sites and is expected to become operational with all of the FAA's commissioned ASOSs.

Automated reports can and should be supplemented with radar products. The system is now virtually complete and covers most of the country. Radar can determine the existence of rain, thunderstorms, tornadoes, snow, and hail. Radar, along with satellite imagery, is better at determining the extent of phenomena than either a manual or automated observation.

Fronts

When flying through a front, a pilot should be prepared for the changes in flight conditions from one air mass to the other. This is sometimes quite abrupt. Abrupt changes indicate a narrow frontal zone. At other times, the changes are very gradual, indicating a broad, weak, or diffuse frontal zone. Pilots should anticipate changes in cloud cover, temperature, moisture, wind, and pressure when penetrating a frontal zone. Frontal systems produce every weather hazard so far discussed.

Essentially pilots have two options when dealing with frontal systems: penetrate or avoid. If a pilot elects to penetrate, there are several additional options. These options are directly related to pilot and aircraft capabilities. A pilot may elect to fly over the front, penetrate at high or low altitude, or attempt to fly under the front. The pilot may want to fly through the front as close to a 90° angle as possible to reduce exposure, or accept the consequences and fly parallel to the frontal band.

> **CASE STUDY**
> Our route from Tulsa, Oklahoma, to Springfield, Illinois, was dominated by a weak cold front. Weather reports and forecasts indicated relatively clear conditions on both ends of the route and tops to about 6000 ft. Beneath the clouds, conditions were turbulent, with marginal ceilings and visibilities. The decision was clear: With an instrument-rated and current pilot, and an aircraft with instrument capability, VFR over the top was acceptable. We filed VFR and the flight was completed without incident.

Should flight on top be tried by a non-instrument-rated pilot? Well, you have to evaluate the risk. An engine failure could lead to disaster. Navigational error or instrument malfunction could lead to getting caught on top. If this should occur, it cannot be overemphasized that the pilot obtain assistance from ATC as soon as possible.

Depending on the type of front and associated weather, a pilot may elect to penetrate the front.

A pilot may wish to avoid frontal weather by either flying around the front or waiting for frontal passage. Flight decisions must be based

> **CASE STUDY**
> A stable Pacific cold front was along our route from Van Nuys to San Francisco. We were flying a Cessna 172. Thunderstorms were neither reported nor forecast. The freezing level was forecast to be around 9000 ft in the south and 6000 ft in the north. We flight-planned along the coastal route, where minimum altitudes were 6000 ft or lower. Under such circumstances, flight over the coastal mountains into California's Central Valley, with minimum altitudes around 10,000 ft, was out of the question.

on the capability of the pilot and aircraft as well as on the weather. In the case of frontal thunderstorms the flight decision is easy—NO GO!

There are really only two ways to avoid or reduce the exposure to frontal turbulence: Fly above the front or penetrate the front as close to perpendicular as possible. Penetrating a front at a 90° angle may work for a cold or stationary front, but may not be satisfactory for a warm or occluded front because of the extensive areas of frontal weather associated with these fronts.

Icing problems exist mostly with winter fronts. With flight altitudes often governed by terrain clearance, freezing level heights are a real concern in flight planning. Additionally, in the mountains and high plateaus the freezing level is often at the surface. In the summer, high freezing levels may create icing conditions into the lower flight levels.

Usually the solution to the icing problem is to fly high or low. Fly above the area of potential icing or below the freezing level. This might not be possible in aircraft with marginal performance, especially in the western United States with its high minimum en route altitudes. Another option is to avoid visible precipitation. That is, remain clear of clouds and areas of precipitation when temperatures are less than 0°C.

With strong cold fronts, ceilings and visibilities are generally good, except in areas of heavy showers. Fog is unlikely because of gustiness and strong winds. A VFR pilot might elect to fly under the front. The pilot could expect a relatively narrow band of, sometimes heavy, precipitation. The pilot could expect a relatively narrow area of low ceilings and visibility. With an active front accompanied by thunderstorms the only option is to wait on the ground. However, with care, a weak front may be negotiated.

CASE STUDY

We had remained overnight in Huntington, West Virginia. Our next leg was to Cincinnati, Ohio, to repair the radio. A weak cold front was moving through the area. Without instrument flight capability, going through or over the front was not an option. As usual with a weak front, ceilings and visibilities were poor, and because of the large frontal boundary, frontal passage was difficult to determine. (This was before the days of satellite imagery.) We knew the weather in Cincinnati was clear. After departure we followed the river (IFR: I follow rivers) and had to negotiate about 10 miles of poor weather before getting into the clear. This can be potentially dangerous flying at low altitudes in poor visibilities, with all the cables and catenaries crossing the river. Exiting the frontal boundary the weather was scattered stratocumulus, with unrestricted visibilities.

With warm fronts, VFR pilots will have to contend with widespread areas of low ceilings and visibilities, and the possibility of freezing precipitation. Pilots must use caution under these circumstances and have a suitable alternate airport close at hand during the entire flight.

Weather and flight conditions in an occluded front will have characteristics of both cold and warm fronts.

With stationary fronts, flight is generally smooth, except when cumuliform clouds are present. Low clouds, ceilings, and visibilities may persist.

Some examples of satellite images of frontal boundaries were presented in Chap. 5. The satellite image is an excellent product to make an initial determination on the strength of a frontal system.

A review of Figs. 6-6 and 6-7 provides another example. Recall that the Area Forecast reports a weak frontal boundary and that the forecast south of a Charleston to Colliers line and the remainder of South Carolina indicates no significant weather. The satellite images indicated clear skies for most of South Carolina.

The satellite images in Figs. 6-6 and 6-7 show clouds along the frontal boundary in southern Georgia and Alabama, and the Florida panhandle. However, these clouds are not as bright as those in North Carolina and Virginia on either the visible or IR image. Thus, this cloud band is not as thick or high as those where the severe weather is occurring. A review of the Radar Summary Chart in Fig. 6-8 confirms this by indicating only light to moderate rain and rain showers accompanying this band of clouds. The Radar Summary also confirms the lower tops in this area, as compared with those in North Carolina and Virginia.

The following case study was obtained from a NASA Aviation Safety Reporting System report.

This pilot extrapolated between weather reporting stations. As the pilot discovered, conditions between reporting points, especially in mountainous terrain, can be considerably different. There is no mention of the pilot obtaining a standard weather briefing prior to the flight. Often air taxi and air carrier pilots rely strictly on their dispatch

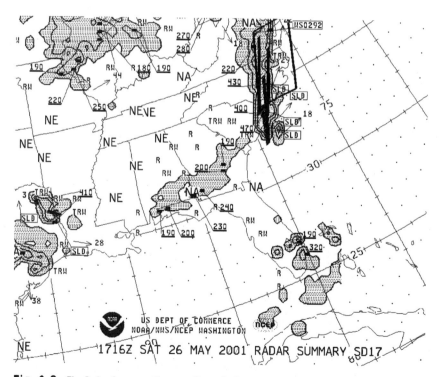

Fig. 6-8. The Radar Summary Chart confirms the lower tops in southern Georgia, Alabama, and the Florida panhandle as compared with those in North Carolina and Virginia.

CASE STUDY

A fast-moving cold front, with a small amount of moisture, moved through east Tennessee during the early morning hours, leaving visibility of 4 to 10 miles, with scattered clouds at 2000 ft, broken clouds at 2000 to 3000 ft, and overcast layers at 6000 to 7000 ft AGL. These conditions were reported between two airports 18 miles apart. Both airports are VFR only, with no instrument approaches.

Between the two airports are two parallel mountain ridges. IFR flights to Mountain City are allowed to descend to 6700 ft for terrain clearance. Our company procedure allows for VFR flight to the airport when reported AWOS weather is VFR. The procedure is to request an IFR descent to Tri Cities airport to an altitude of 3700 ft in order to get below the clouds and then, with good visibility, fly through a mountain pass over a lake to land at Mountain City Johnson County airport. Just prior to takeoff, the AWOS reported Mountain City 2700 ft scattered, 3200 ft broken, 4000 ft overcast, with 4-mile visibility. Elizabethton reported 1700 ft scattered, 2300 ft broken, 6000 ft overcast, with 10-mile visibility. Tri Cities airport reported 1300 ft scattered, 3500 ft broken, 8000 ft overcast, with 7-mile visibility.

I relied on AWOS reports from two airports 18 miles apart, but the actual conditions between the mountain ridges were different. The clouds between the mountains were lowering in the valley. My first officer was new to the company and had never flown to Mountain City. He had been a commuter pilot and was uncomfortable in these types of conditions.

Center cleared us to 6700 ft and told us to change to Tri Cities approach. I told the first officer to ask for radar vectors to Elizabethton, so that we could get below the overcast and get in VFR conditions. We would proceed to Mountain City by way of Elizabethton. Tri Cities approach told us to make a 360° turn to the right and pick up a heading of 090° and descend to 3700 ft. During the turn, I asked the first officer again for Elizabethton AWOS update. He was having difficulty finding the correct frequency. Coming out of the turn to 090, we broke out of the clouds at about 4500 ft. We were flying toward Holston Mountain, level at 3700 ft in what were VFR conditions, so I told the first officer to tell Tri Cities approach that we were in VFR conditions and could proceed to Elizabethton. Tri Cities approach cleared us to Elizabethton. I turned right and paralleled Holston Mountain on my left. I noticed that the overcast was broken, with layers of scattered to broken cloud. I could see Johnson City ahead. Visibility was probably 10 miles or greater. At the southern base of Holston Mountain I saw Elizabethton airport with a good 4000 to 4500 ft broken bases. I instructed the first officer to cancel IFR. We would proceed to Mountain City VFR.

Tri Cities approach told us a Cessna 172 had taken off from Elizabethton, climbing below us, heading west, and that it should be no factor. We were at 3500 ft, descending to 2500 ft. I told the first officer that if the conditions were not favorable, we would make a 180° turn and go back and land at Elizabethton. He agreed that we could climb if necessary.

As we got closer, I saw that the clouds were lower than expected, too low to safely continue. I immediately made a 180° turn and found myself in the clouds. I was too low to talk to approach and had to climb to get back on top and re-establish radio communications. I called Tri Cities approach for an IFR clearance. They would not issue the clearance because the Cessna 172 was about 5400 ft. We were at 4000 ft and climbing. We broke out of the clouds about 5000 ft. My first officer reported a visual sighting of the Cessna 172. I continued VFR at altitudes between 5500 and 6500 ft to stay out of the clouds. Tri Cities gave us a heading of 020°. I saw Elizabethton about 5 miles ahead. We flew over the airport and landed without further incident.

personnel to make the go–no go decision. There is also no mention of the pilot checking with Flight Watch or FSS for updated reports, forecasts, and PIREPs. It does not appear the pilot used all available resources.

The pilot correctly deduced that both airports "...are VFR." Categorically they were reporting "marginal VFR." These facts should have alerted the pilot that conditions between reporting points could be less than that reported at either station, and possibly below basic VFR. However, as the pilot stated: "...company procedure allows for VFR flight to the airport when reported AWOS weather is VFR...."

The pilot descended, cancelled IFR, and proceeded VFR toward the destination. The flight crew decided if conditions were "...not favorable..." they could make a 180° turn and go back and land at Elizabethton. From the report, it appears the crew decided they could climb through the overcast, if necessary. This action, as it turned out, involved a violation of regulations.

Conditions deteriorated, and although the pilot made a 180° turn, they ended up in the clouds. It appears the crew waited too long before attempting to extricate themselves from this situation. Now their predicament was compounded by being below radio coverage. Once they were in contact with ATC a clearance could not be issued because of traffic. Fortunately for this crew, they were able to break out on top.

In the report, the pilot concluded that the weather was too low to attempt such a flight. The pilot also stated: "It was poor judgment to cancel IFR, thus eliminating communications and radar contact...," and that company has since changed it policy for such operations.

Nonfrontal Weather Systems

VFR pilots must be careful dealing with upper-level weather systems, especially in the absence of a front. Typically, weather conditions will be VFR to marginal VFR at the surface. In fact, pilots are often able to operate VFR beneath ceiling with relatively good visibilities, except in the vicinity of showers, during these conditions. Here's the rub. In mountainous areas, there is almost always mountain obscuration. This prevents or precludes VFR operation in these areas. Pilots have to be especially careful not to be caught in box canyons. IFR pilots will typically have to contend with low freezing levels, turbulence, and thunderstorms—often embedded. To safely operate in these areas, the aircraft must have sufficient performance, and ice protection and thunderstorm avoidance equipment. If you don't, have all of that, don't go!

This scenario is illustrated in Fig. 6-9. (The visible image is on the left and the IR image on the right.) Notice that California's Central Valley is mostly clear, with the Coastal and Sierra Nevada Mountains cloud-covered—expect the mountains to be obscured in clouds and precipitation. Visibility in the valleys is good, except in the vicinity of rain and snow showers. Since it's winter and the circulation is cyclonic, bringing in cold air, freezing levels are relatively low. With the moderate winds from such a system, expect moderate mechanical turbulence over the mountains; turbulence may be severe in the vicinity of convective activity. Thunderstorms may be embedded in the bands of showers, for example, the thick band with high, cold tops moving into northern California.

Some upper-level weather systems develop clouds and weather in bands (like the system in Fig. 6-9; note the absence of a frontal boundary). Typically, aviation forecasts cannot take this phenomenon into account. Therefore, the forecaster will cover the area with a "broad brush" approach. A pilot, on seeing a clear area, may assume the weather has passed. However, the weather then deteriorates with the approach of the next band. With closed lows and hurricanes, expect weather in bands. The satellite picture will often confirm or refute the existence of bands.

Another factor associated with upper-level lows and troughs is an unstable air mass. The low or trough may move through an area, bringing clear skies behind. However, with surface heating, abundant moisture, and an unstable air mass, guess what comes next. You've got it—thunderstorms! Usually the forecast has this pretty well in hand, although by now we should be able to anticipate such phenomena. Along with radar, the satellite image can be a significant help in identifying the exact location of convective weather. Several examples were presented in Chap. 5.

Besides radar and satellite data, other sources of severe weather information are contained in Convective SIGMETs (WST), Alert Weather Watches (AWW), and to a lesser degree the Convective Outlook (AC). (The Convective Outlook is an outlook product for advanced planning only.) These products provide information on the atmosphere that relates to severe weather. All of the products contain forecast discussions. They provide additional insight into the overall weather picture. A few words of caution: FSS briefers will not normally provide the outlook portion of WSTs—which contain the discussions—except on request. The Convective Outlook is just that, an outlook. The AC can never be used as a substitute for appropriate aviation weather forecasts (Convective SIGMETs, Area Forecasts, Terminal Forecasts).

The jet stream has its hazards of turbulence and strong winds. We have already discussed their significance and strategies. Hazards associated with vorticity are those produced by vertical motion mechanisms. The mention of these phenomena in forecasts or their

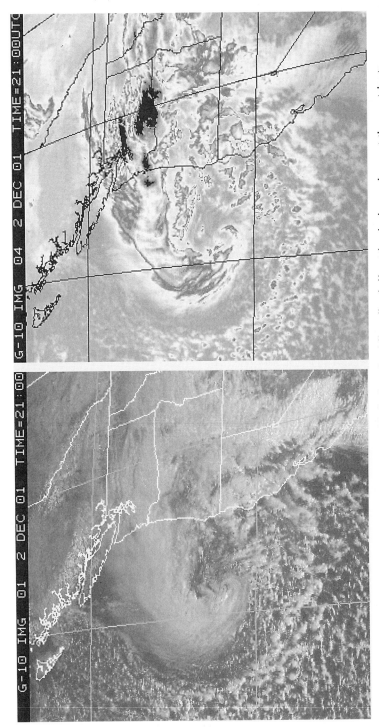

Fig. 6-9. Notice that California's Central Valley is mostly clear, with the Coastal and Sierra Nevada Mountains cloud-covered—expect the mountains to be obscured in clouds and precipitation.

appearance on charts should alert us to the fact that vertical motion is, or will be, occurring. En route, pilots should watch for unforecast wind speeds or shifts or temperature changes. These can signal a change in the weather pattern, which should cue the pilot to obtain additional information, such as current weather reports and revised forecasts—not to mention that a PIREP will alert forecasters, briefers, and other pilots to a potentially hazardous, unforecast change in the weather.

Like thunderstorms, the key to the hurricane hazard is avoidance. Thunderstorms tops associated with tropical cyclones frequently exceed 50,000 ft. Winds in a typical hurricane are strongest at low levels, decreasing with altitude. However, winds in excess of 100 knots at 18,000 ft are not uncommon. Severe to extreme turbulence and severe low-level wind shear are to be expected. Turbulence increases in intensity in spiral rain bands and becomes most violent in the wall cloud surrounding the eye. And severe icing can be expected above the freezing level, which may be in the lower flight levels. Altimeter errors, because of the extreme low pressure, may be as much as 2000 ft. Flying to the right of the storm, a pilot can take advantage of a tailwind; to the left the pilot will encounter the strongest headwinds, increasing fuel consumption and prolonging the risk.

The crew of the record-setting Voyager round-the-world flight in 1986 used this technique: In the western Pacific, off the Philippine Islands, they altered their flight plan to take advantage of the tailwinds north of a tropical storm. They had to be careful because the airplane was not constructed to withstand a large amount of turbulence. Satellite imagery played a large part of this game plan.

CHAPTER 7

Flight Planning Strategies

A t this point, we should have a sound understanding of satellite and radar products. Armed with this knowledge, we can move on to flight planning and cockpit strategies. As we'll see, these subjects are often interrelated. Therefore, even though the focus of this chapter is primarily flight planning, we occasionally bring in topics that apply to en route operations.

We'll begin with flight planning. Flight planning starts with an appraisal of personal minimums, and consists of evaluating all elements physically related to the flight, such as terrain, altitudes, and the environment. With an evaluation of terrain and altitudes complete, we can move on to an assessment of the environment—weather, personal minimums, and alternates. The first part of the environmental evaluation is obtaining the "complete picture" through a preflight weather briefing. This knowledge is then applied to our personal minimums and available alternates. An assessment of terrain, altitudes, and environment is the first step in the go–no go decision.

Unlike being a "little pregnant," flying through weather offers some middle ground. For example, we can plan the flight in stages, landing short of our ultimate destination. Then we can take another look at the weather. But there are two caveats to this option. First, we must know when to abandon the plan. When it's not meant to be, it's not meant to be! We must know when to call it a day. Second, we must have an alternate plan or two (Plan B, Plan C, Plan *n*)—more about this later in the chapter and in Chap. 8, "Evaluating Risk." Below is an example of such a situation.

We were on a flight from Amarillo, Texas, to Albuquerque, New Mexico. The weather was good through Tucumcari, but deteriorated between Tucumcari and Albuquerque. Passing Tucumcari, I checked with Flight Watch and received the bad news. The weather ahead was IFR to MVFR. The plan was to fly direct, via the Anton Chico and Otto VORs. With the deteriorating weather ahead, I decided to go IFR—I follow roads. With low ceilings and visibilities, and little in way of landmarks, the safest option was to follow I-40. The terrain and clouds merged about 20 miles west of Santa Rosa, New Mexico. It was afternoon and we had been flying for about 4 hours. With night approaching, poor weather, and fatigue a factor, the only viable option was to return and land at Santa Rosa.

Back to flight planning. With a go decision, our next step is preparation. In addition to our aircraft, its airframe, engine, and equipment, we must consider ourselves. Are we fit for flight? This means certified, current, and physically and mentally prepared. Don't forget the passengers; are they physically prepared for the flight? What if it's cold? What would happen in the event of an emergency landing? Are we prepared to survive a day or two in the open? We'll talk about this and other survival issues later.

Both of these case studies illustrate the need to obtain all available information. If all other airports in a general location are reporting

CASE STUDIES

The pilot tuned in the AWOS and received a report of 100 scattered. Over the final approach fix, the pilot experienced a solid undercast with tops at 1300 ft. There were no suitable VFR airports in the area. The pilot, unable to land, was forced to return to the departure airport.

In another case, the AWOS was reporting clear. On approach an air carrier stated: "Unable to make a visual approach because of clouds. The airport is the only clear spot in the area."

overcast, the pilot should view a report of "100 scattered" with some skepticism. On the other hand, the airport being the only clear spot in the area is not uncommon. If we have the capability of flying IFR, stay on the IFR flight plan until landing VFR is assured. We don't know if the satellite picture, PIREPs, or forecasts were obtained by the reporters. These may have given the pilots additional insight.

The reporter in the first incident commented: "This AWOS error could have caused a serious fuel problem for a long-range flight arriving with minimum required fuel remaining." For IFR operations, regulations take these types of problems into consideration by requiring fuel to alternate airports, plus an additional fuel reserve. Minimum fuel reserves for VFR flights are just that—minimum. Minimum does not necessarily equate to safe. As soon as an undercast is encountered, a VFR pilot must check to determine its extent. It makes no sense to fly for 45 minutes over an undercast betting your life that an airport will be clear with only a 30-minute fuel reserve.

It might be a case of a faulty sensor. When it verifies this condition, the FSS will issue a NOTAM indicating an individual sensor or sensors are unreliable. The FSS or controlling agency will also notify maintenance personnel. As in case of the contract observer failing to correct a nonrepresentative report, we as pilots have an obligation to report faulty equipment.

Personal Minimums

Periodically, we will talk about minimums. Minimums are just that—minimums. Minimums do not necessarily mean safe. Each pilot, whether VFR or IFR, needs to set his or her own safe minimums. For example, it makes no sense to depart VFR with an overcast of 500 ft, visibility 1 mile in a mountainous area. It may be technically legal, but not safe! Nor does it make any sense to fly in an area of potential icing without ice protection equipment or in an area of thunderstorms in an aircraft without storm detection equipment. Again, legal, but not safe! Minimums, and the decision to initiate a flight, should be based on a pilot's training and experience, and the aircraft and equipment to be flown.

During the last 10 years, approximately 27 percent of all general aviation airplane accidents involved adverse weather conditions. Of these, 30 percent caused fatalities. The biggest causal factor, as in the past, involved continued VFR (Visual Flight Rules) flight into instrument meteorological conditions (IMC). This accounts for 56 percent of fatalities for all weather-related accidents. By a wide margin, the most dangerous situation is VFR flight into IMC!

What are your personal minimums? In our Air Force Aero Club, student pilots were limited to flight in 10 knots or less of surface wind, with no more than a 5-knot crosswind component. As a new flight instructor I imposed specific limits based on the individual student's training and experience. For solo cross-country flights, students always had an alternate in case adverse winds developed.

Student pilots are guided through their initial training under the direct supervision of a flight instructor. Military pilots are shepherded by more experienced pilots. And, ironically, the airlines have an extensive dispatch system to ensure the safety of their flights. However, after initial certification there is no such safety umbrella for the general aviation pilot. This fact can be directly related to the accident record. In a very real sense we hold our fate in our own hands. A new certificate or rating should be considered a "license to learn."

When we talk about personal minimums, there are a number of factors to consider. These include:

- Training
- Experience
- Currency
- Aircraft
- Weather
- Time of day
- Terrain
- Physical condition
- Psychological condition

Notice that there are a number of considerations that are not neces-sarily weather related. Nevertheless a pilot must take them into con-sideration as part of the flight planning strategies.

As our levels of training and experience increase, we may wish to consider different minimums. (We'll discuss this in more detail in the following section.) As a flight instructor, I tailored student minimums to their training and experience. For example, I had a student flying out of Lancaster's Fox Field, in California's Mojave Desert. We trained in strong, gusty surface winds. When the student was proficient I would increase the minimums. Some pilots obtain an instrument rating without ever having flown in the clouds. Do they have the experience to operate in actual instrument conditions? A prudent pilot would have another qualified, experienced pilot or flight instructor along until he or she became familiar with flight in actual instrument conditions.

Currency with the type of operation is another personal minimum factor. Here again, legal does not necessarily mean safe. If we've been recently checked out in a complex or high-performance airplane or qualified to fly at night, we would certainly want to gain experience before tackling weather close to minimums, either VFR or IFR. Night operations present additional challenges and risks.

We've touched on the concept of the "complete picture." Notice throughout this discussion we continually refer to other aviation weather products for additional information and verification. Satellite imagery must be used in conjunction with the surface observations, and the Surface Analysis, Weather Depiction, and Radar Summary Charts. Satellite imagery can often reveal the extent of weather phenomena, such as stratus and fog layers, location and intensity of some types of fronts and convective activity, and regions of snow or clear skies. The combined application of satellite and observational products can confirm or refute the accuracy of both manual and automated weather observations.

Weather advisory phenomena usually lie well within their desig-nated boundaries. Satellite imagery can often further delineate these areas. Additionally, imagery may expose the fact—sorry I couldn't help it—that weather phenomena are developing or dissipating as forecast. Recall Fig. 2-18; this would seem to be a good day for VFR flight along the coast and in the mountains. In California's Central Valley even IFR flights may have difficulty with extensive areas of near zero-zero conditions.

The best use of ground radar information is to isolate general areas of coverage of echoes. You must avoid individual storms from inflight observations either by visually sighting or by airborne radar. It is better to avoid the whole thunderstorm area than to detour around individual storms unless they are scattered.

Thunderstorms build and dissipate rapidly. Therefore, do not attempt to plan a course between echoes.

There are a number of advantages to a self-briefing. For the weatherwise pilot, experienced in accessing and decoding the hieroglyphics of aviation weather, it's often easier to get the "complete picture" of the weather, as compared to talking with an FSS controller. This is not to criticize; it's just that FSS controllers are trained to distill the weather picture for pilots, and some pilots would rather see the "big picture" in its entirety for themselves. Computer access to weather information gives pilots all the raw data, often helpful in answering tough go–no go decisions. Another advantage is that, assuming you can log into the Internet without being "put on hold," it's often quicker to fire up the computer than it is to reach a live FSS person on the telephone. This is especially true during peak flight planning times and during bad weather. Access to color graphics is another distinct plus with most self-briefing services.

With a knowledge of the three-dimensional atmosphere, pilots should be able to put together the whole, or "complete picture." By understanding why some weather systems are benign and other severe, and how they are modified, then integrating this knowledge with the preflight weather briefing and updates en route, pilots should be able to make intelligent, safe weather decisions. But this is only half of the safety equation. Most frequently, an accident occurred because the pilot failed to obtain complete and accurate information, attempted to exceed aircraft performance, or simply continued flight beyond their capability or below safe minimums. Virtually all were preventable!

Thus far we've related terrain to specific aviation weather hazards and flight situations. In addition to weather conditions, pilots must include aircraft performance in their decision. Aircraft performance and equipment refer to density altitude, service ceiling, availability of supplemental oxygen, and ice protection and thunderstorm avoidance equipment. If the aircraft doesn't have the performance or equipment, don't go! Other factors would be alternate landing sites and time of day.

A pilot's physical and psychological state must also be considered. Proper rest and good mental condition are paramount. It makes no sense for a tired, hungover pilot to attempt to take on weather, or any flying for that matter. Fatigue is a prime example. It doesn't make any sense for a pilot suffering from fatigue to attempt a night landing to instrument minimums. In fact, this can be a recipe for disaster. Pilots need to set their own minimums based on these factors.

Figure 7-1 contains a table of suggested minimums based on pilot qualification and type of operation. [If the FAA put this together it would be called a "matrix." So what's the difference between a table and a matrix? The cost, of course. The FAA would grant a contract to a university or corporation worth tens of thousands of dollars to come up with such a matrix. Your tax dollars at work(?).]

Refer to Fig. 7-1. The left column describes pilot qualification (STU—student, PVT—private, COM—commercial, INST—instrument). Two additional categories are dual flight instruction VFR (DUAL V) and IFR (DUAL I). Across the top are operational categories. Operational categories are subdivided into DAY and NIGHT, along with winds aloft limits for cross-country operations, and surface wind limitations. Note that student night solo is not authorized (NA), even though it is permitted by regulations with specific training and an instructor's endorsement. Ceiling and visibility are given in feet and statute miles (7500/7—ceiling 7500 feet and visibility 7 statute miles). For dual flights, basic 14 CFR 91 limitations are authorized (FAR); with sustained surface wind and gust left to the discretion of the instructor pilot (PD—pilot discretion). For commercial pilots and dual flights, maximum demonstrated cross-wind component as listed in the *Pilot's Operating Handbook* (POH) are allowed.

Before we continue, let's concede that the minimums described in Fig. 7-1 are not hard and fast. Because of training and experience, an instructor may wish to raise or lower a student's solo limitations.

PERSONAL MINIMUMS

PILOT	CROSS COUNTRY		SURFACE WIND				LOCAL		PATTERN	
	DAY	NIGHT	WINDS ALOFT	CROSS WIND	SUS-TAINED	GUSTS	DAY	NIGHT	DAY	NIGHT
STU	7500/7	NA	25 KT	7 KT	15 KT	NONE	5000/5	NA	2000/3	NA
PVT	7500/5	7500/5	25 KT	10 KT	20 KT	5 KT	4000/5	4000/5	2000/3	2000/3
COM	7500/3	7500/3	35 KT	POH	25 KT	10 KT	4000/3	4000/3	2000/3	2000/3
DUAL V	FAR	4000/5	35 KT	POH	PD	PD	FAR	4000/3	FAR	1500/3
DUAL I	FAR	800/2	35 KT	POH	PD	PD				
INST	500/1[1]	1000/2								

[1] Or, FAA published Takeoff and IFR Departure minimums, including climb gradients, whichever is greater.

Fig. 7-1. Personal minimums should be based on training and experience, along with a knowledge of the aircraft and your physical and psychological condition.

A private pilot with years of experience and thousands of hours in a specific make and model of aircraft may be competent to exercise limits in the commercial or dual categories. We know that maximum demonstrated cross-wind component is not an absolute limit. Pilots should consider the personal minimums in Fig. 7-1 as a starting point. Also note that the relatively high local and cross-country ceilings are due to the high terrain in the western United States. In the midwest and along the east coast, in the absence of mountainous terrain, these would normally be lower. Our United States Air Force Aero Club in England had ceiling minimums of 2500 ft because of the flat terrain. This was certainly sufficient for operations in eastern England.

Notice in Fig. 7-1 that night minimums are typically higher than day. This is a direct reflection of the additional hazards of night flight. Instrument minimums are also typically higher than those specified in the regulations. These minimums were developed for single-engine and light twin-engine airplanes. Why? Even though a single-engine pilot, operating under 14 CFR 91, is not directly prohibited from making a zero-zero takeoff, it isn't safe. Now consider that, in the event of an engine failure, the pilot is still required to be able to: "make an emergency landing without undue hazard to persons or property on the surface." Also, recall that: "No person may operate an aircraft in a careless of reckless manner so as to endanger the life or property of another."

Figure 7-1 does not address personal fuel minimums. The following fuel reserves are recommended and should be seriously considered.

- Minimum 1-hour reserve for all flights

- Minimum 2 hours of fuel before takeoff for pattern or local flights

- Full fuel for cross-country and IFR flights

On certain airplanes, filling the tanks to the tabs may be satisfactory and allow the accommodation of additional passengers and baggage. Most four-place airplanes are designed to accommodate a full passenger load with partial fuel, or fewer passengers and baggage with a full fuel load, but not both! Pilots must understand the limitations of their aircraft.

The Weather Briefing

In 1983 the FAA established three types of briefings. These continue today:

- Standard briefing

- Abbreviated briefing

- Outlook briefing

- Inflight briefings

Federal aviation regulations require each pilot in command, before beginning a flight, to become familiar with all available information concerning that flight. This information must include: "For a flight under IFR or a flight not in the vicinity of an airport, weather reports and forecasts…and any known traffic delays of which the pilot has been advised by ATC."

Additional regulations specify fuel and alternate airport requirements. The regulations do not, however, require that meteorological and aeronautical information be obtained from the FAA. In the early 1990s, the FAA contracted with two organizations to provide meteorological and aeronautical information to be provided by computer, known as a Direct User Access Terminal, or DUAT. Since then, a number of other vendors provide pilots with this information. The FAA is in the process of certifying vendors as official sources of weather information. Pilots must be careful to ensure their weather provider's products constitute a "legal" briefing.

The FAA's standard briefing is designed for a pilot's initial weather rundown prior to departure. Standard briefings are not normally provided when the departure time is beyond 6 hours, nor current weather beyond 2 hours. It is to the pilot's advantage to obtain a standard briefing, or update the briefing, as close to departure time as possible.

Before beginning a briefing, the FSS controller must obtain background information that is pertinent and not evident or already known. The amount of information varies with the training and experience of the briefer, weather conditions, and the pilot's request. Pilots can assist the briefer and reduce delays by volunteering the following information.

1. *The type of flight planned.* Always advise the briefer if the flight can be conducted only by VFR, or that an IFR flight is planned, or a flight that can be conducted IFR. Normally, the briefer will assume a pilot is planning VFR, unless stated otherwise. Student pilots should always identify themselves to help the briefer provide a briefing tailored for a student's needs. Also, new or low-time pilots and pilots unfamiliar with the area will receive better service if they advise the briefer. This alerts the briefer to proceed more slowly, with greater detail.

2. *The aircraft number or pilot's name.* This is evidence that a briefing was obtained, as well as an indicator of FSS activity. In the absence of an aircraft number, the pilot's name is sufficient. Most briefings are recorded and reviewed in case of incident or accident; it's in the pilot's interest to get "on the record" as having received a briefing.

3. *The aircraft type.* Low-, medium-, and high-altitude flights present different briefing problems. This information allows briefers to tailor the briefing to a pilot's specific needs. By knowing the aircraft type, the briefer, many times, can estimate general performance characteristics such as altitude, range, and time en route.

4. *The departure airport.* Pilots must be specific; they know the airport, but the briefer usually doesn't. This is important with FSS consolidation, 800 phone numbers, and in metropolitan areas.

5. *The estimated time of departure.* The estimated time of departure is essential, even if general, such as morning or afternoon.

6. *The proposed altitude or altitude range.* This information is needed to provide wind- and temperature-aloft forecasts. If an altitude range is specified, for example 8000 to 12,000 ft, the briefer can provide the most efficient altitude for direction of flight.

7. *The route of flight.* The briefer will assume a pilot is planning a direct flight, unless otherwise stated. If not, a pilot must provide the exact route or preferred route, and any planned stops. This will assist the briefer in providing weather for the planned route.

8. *The destination airport.* Again, pilots must be specific. If not, a pilot might not receive all available weather and NOTAM information.

9. *Estimated time en route.* Many briefers can estimate time en route on the basis of aircraft type. This information is needed to provide en route and destination forecasts. Total time en route is essential when stops or anything other than a direct flight are planned; for IFR flights, the estimated time of arrival is required to determine alternate requirements.

10. *Alternate airport.* If you already have an alternate in mind, provide it at this time. FSS equipment will automatically display alternate airport current weather, forecast, and NOTAMs to the briefer.

This might seem like a lot of information, but it really isn't. The briefer must obtain this information before or during the briefing. Providing background information will allow briefers to do a better job, which is to provide a pilot with a clear, concise, well-organized briefing, tailored to the pilot's specific needs.

Alright, the background information has been provided; what can a pilot expect in return? The briefer is required, using all available weather and aeronautical information, to provide a briefing in the following order. Pilots should be as familiar with this format as the mnemonic CIGAR (controls, instruments, gas, attitude, runup), or the IFR clearance format.

1. *Adverse conditions.* Any information, aeronautical or meteorological, that might influence the pilot to cancel, alter, or postpone the flight will be provided at this time. Items will consist of weather advisories, major NAVAID outages, runway or airport closures, or any other hazardous conditions.

   The adverse conditions provided should be only those pertinent to the intended flight. This is one reason why the pilot must provide the briefer with accurate and specific background information. The briefer should then furnish only those conditions that affect the flight. There is, unfortunately, some paranoia among briefers where they provide anything within 200 miles of the flight, whether it's applicable or not.

2. *VFR flight is not recommended (VNR)*. Undoubtedly the VNR statement is the most controversial element of the briefing, nevertheless, the FAA requires the briefer to: "Include this recommendation when VFR flight is proposed and sky conditions or visibilities are present or forecast, surface or aloft, that in (the judgment of the specialist), would make flight under visual flight rules doubtful."

The following case study is from a NASA Aviation Safety Reporting System report.

> **CASE STUDY**
>
> A call to the FSS indicated the weather had started to deteriorate north of Lexington, Kentucky, with lower ceilings and visibility developing. The briefer stated that VFR flight was not recommended. Dayton and Fort Wayne were at best marginal VFR. A front was moving in from the west. I elected to proceed. Somewhere south of the Falmouth VOR lower ceilings were encountered. I flew through some clouds to VFR on top right around 6500 ft. I called ATC prior to the encounter. Dayton approach called Fort Wayne to check for possible holes or VFR airports in the area. None were VFR. I had approximately 2 hours and 30 minutes of fuel left. There was still time to head south to VFR airports with London a little over an hour south.
>
> Why not go back? Dayton approach advised me I'd broken FAR 91. "What do you want to do?" Dayton approach asked. I replied that I'd like to go to Versailles because it had an ADF and believe I also said I could land at Dayton. Dayton approach told me they could help only if I declared an emergency. This tells me that I am on my own unless I declared an emergency. I elected to declare an emergency to obtain radar assistance and separation from other aircraft. The controller provided assistance to Dayton.

The pilot received a weather briefing that indicated a front was moving in from the west producing low ceilings and visibilities. The pilot was advised by the briefer that: "VFR flight is not recommended." The pilot elected to proceed. The issuance of VNR should not in itself cause a pilot to cancel a flight, but it does indicate a need to take a very close look at the weather situation.

The pilot encountered clouds, and rather than retreat, climbed through the clouds. After contacting ATC, approach control checked for possible VFR airports in the area. Although center, approach control, and tower personnel can be helpful with a weather encounter, that is not their primary function and they have relatively limited resources in this area. Pilots should use the FSS, preferably Flight Watch, to assess the situation well before weather deteriorates beyond the pilot's qualifications or capability.

Then, the report indicates that the pilot and ATC got into a "spitting" contest. The pilot states that ATC refused help unless the pilot "declared an emergency." The pilot then states: "I elected to declare an emergency to obtain radar assistance...." At this point the controller provided assistance to Dayton. In fact, an emergency can be declared by any of three entities: the pilot, air traffic control, or the person responsible for the operation of the aircraft. Although it is difficult to determine the exact circumstances out of context and with the limited information available, the pilot should have declared an emergency when it became evident that less than VFR conditions would be encountered. The controller should have handled this situation as an emergency, whether it was declared by the pilot or not.

Safely on the ground the pilot spoke with ATC on the phone. The pilot states that when the emergency was declared, the watch supervisor took over—"I didn't have much choice," the watch supervisor said. The pilot further says the watch supervisor said, "I had done a good job and didn't get rattled as most VFR pilots would have in similar situations." As a former FSS supervisor, active Aviation Safety Counselor, and flight instructor, this is not the behavior we want to compliment or encourage. In my opinion this was an inappropriate comment by the controller.

The pilot further comments the decision to continue was incorrect and that even to start the flight in the first place was a mistake. So how do we train pilots not to exceed their ability or that of their aircraft and discontinue a flight when the weather goes sour?

3. *Synopsis.* The synopsis is extracted and summarized from FA and TWEB route synopses, weather advisories, and surface and upper-level weather charts. This element might be combined with adverse conditions and the VNR statement, in any order, when it would help to more clearly describe conditions.

These three elements should provide the big picture, part of the complete picture. The synopsis should indicate the reason for any adverse conditions, and tie in with current and forecast weather. During this portion of the briefing pay particular attention for clues of icing, even if a weather advisory is not in effect.

4. *Current conditions.* Current weather will be summarized: point of departure, en route, and destination. Relevant PIREPs, the satellite image, and weather radar reports will be included. Weather reports will not normally be read verbatim, and might be omitted if proposed departure time is beyond 2 hours, unless specifically requested by the pilot.

5. *En route forecast.* The en route forecast will be summarized in a logical order (climbout, en route, and destination) from appropriate forecasts (FAs, TWEBs, weather advisories, and prog charts). The briefer will interpret, translate, and summarize expected conditions along the route. Specifically, look for forecast bases and tops.

6. *Destination forecast.* Using the TAF where available, or appropriate portions of the FA or TWEB forecast, the briefer will provide a destination forecast, along with significant changes from 1 hour before until 1 hour after ETA. We're especially interested in wind, visibility, weather, type of precipitation, and cloud bases.

7. *Winds aloft forecast.* The briefer will summarize forecast winds aloft for the proposed route. Normally, temperatures will be provided only on request. Request temperatures aloft. We want to know if we're going to be below, at, or above the freezing level. Temperature at our flight plan altitude is an indicator of icing severity, as well as a warning to consider aircraft performance.

8. *Notices to Airmen (NOTAMs).* The briefer will review and provide applicable NOTAMs for the proposed flight that are on hand, and not already carried in the *Notices to Airmen* publication. Pay particular attention to landing area conditions. We might want to check surface conditions for en route airports as possible alternates. This will require a specific request.

9. *Other services and items provided on request.* At this point in the briefing, briefers will normally inform the pilot of the availability of flight plan, traffic advisory, and Flight Watch services, and request pilot reports. On request, the controller will provide information on military training route (MTR) and military operation area (MOA) activity, review the *Notices to Airmen* publication, check loran or GPS NOTAMs, and provide other information requested.

It's not necessary to copy all the information provided because much is supplementary and provides a background for other portions of the briefing. Pertinent information should be noted, and it's often advantageous to copy this data. There are many forms available. It's often helpful to have a map containing weather advisory plotting points. This chart is contained in Fig. C-1 in Appendix C, "Weather Advisory Plotting Chart/Locations."

Briefers provide abbreviated briefings when a pilot requests specific data, information to update a previous briefing, or supplement an FAA mass dissemination system (Transcribed Weather Broadcast, Telephone Information Briefing Service, or Pilot's Automatic Telephone Weather Answering Service).

When all that's required is specific information, a pilot should state this fact and request an abbreviated briefing. Because the briefer must normally make a request for each individual item, it's extremely helpful to request all items at the beginning of the briefing, thus reducing delays. The briefer will then provide the information requested. When this procedure is used, the responsibility for obtaining all necessary and available information rests with the pilot, not the briefer. Pilots must realize that the briefer is still required to offer adverse conditions. Pilots sometimes become irritated when the briefer mentions weather advisories; however, this is an FSS handbook requirement.

Pilots requesting an update to a previous briefing must provide the time the briefing was received and necessary background information. The briefer will then, to the extent possible, limit the briefing to appreciable changes. An alarming number of pilots, when asked the time of their previous briefing respond, "I got the weather last night." Needless to say, this practice does not comply with regulations. These individuals should be requesting a standard briefing.

When requesting supplemental information for an FAA mass dissemination system, again, the briefer must have enough background information, and the time the recording was obtained. The extent of the briefing will depend on the type of recording and time received.

With a proposed departure time beyond 6 hours, an outlook briefing will normally be provided. The briefing will contain available information applicable to the proposed flight. The detail will depend on the proposed time of departure. The further in the future, the less specific. As a minimum, the outlook will consist of a synopsis and route/destination forecast.

Although the practice is discouraged, unless unavoidable, briefings once airborne will be conducted in accordance with a standard, abbreviated, or outlook briefing as requested by the pilot. As with any briefing, sufficient background information must be made available.

Briefings can be obtained in person, over the telephone, or by radio. The preferred methods are to obtain a weather briefing in person or by phone. Initial briefings by radio are discouraged, except where there is no other means. The reasons are simple. The cabin of an aircraft plunging into the wild gray yonder is no place to plan a flight. Attention must be diverted from flying the aircraft to the briefing. Especially with marginal weather, certain pilots have a tendency to push on, regardless of conditions, not to mention the fact that it usually unnecessarily ties up already congested radio frequencies.

The weather briefing is a cooperative effort between the pilot and the FSS controller. Preliminary planning should be complete, including a general idea of route, terrain, minimum altitudes, and possible alternates. Where available, obtain preliminary weather from one of the recorded services. From the broadcast, determine the type of briefing required—standard, abbreviated, or outlook.

During the briefing, try not to interrupt, unless the briefer is going too fast. Often pilots interrupt with a question that was just about to be answered. This can cause the briefer to lose his or her train of thought, resulting in the inadvertent omission of information.

Briefers make mistakes, and many are not pilots. At the end of the briefing, don't hesitate to ask for clarification or additional information on any point you do not completely understand. If conditions are right for turbulence or icing and these phenomena were not mentioned, ask the briefer to verify that there are no weather advisories.

With this as a background and FSS staffing being further reduced, the question becomes, "How can a pilot best use the services available?"

DUAT, and virtually all other commercial services, use National Weather Service products, which contradicts the misconception that computer briefings somehow provide different products than available through an FSS.

When you use these services, it's essential to know what information is available. Pilots using a commercial system must check with the vendor to determine how their system handles aviation products. Certain advisories, for example, CWAs, might not be available on some systems, and none provide local NOTAMs. Know your service, and check with an FSS for any additional information required, or to clarify anything you don't understand—remember the disclaimer.

There are a number of advantages to self-briefing. For the weather-wise pilot, experienced in accessing and decoding the hieroglyphics of aviation weather, it's often easier to get the "complete picture," as compared to talking with an FSS controller. This is not to criticize; it's just that FSS controllers are trained to distill the weather picture for pilots, and some pilots would rather see the "big picture" in its entirety for themselves. Computer access to weather information gives pilots all the raw data, often helpful in answering tough go–no go decisions. Another advantage is that, assuming you can log into the Internet without being "put on hold," it's often quicker to fire up the computer than it is to reach a live FSS person on the telephone. This is especially true during peak flight planning times and during bad weather. Access to color graphics is another distinct plus with most self-briefing services. With these advantages come the responsibility to decode, translate, interpret, and apply information to a flight. The pilot will have to sift through the mountains of written data, formally reserved for the FSS controller, to determine if a particular flight is feasible under existing and forecast conditions, and aircraft/pilot capability.

On the other hand, many FSS controllers are excellent interpreters and translators of the weather. They are a resource that should not be overlooked. They can, often, provide valuable assistance to the pilot, especially if any information is missing or unclear.

The sheer amount of information might be overwhelming, especially for long-distance flights. A pilot might have to study several pages for a single sentence that applies. Finally, if you have a problem with one of these services you'll have to contact the vendor.

As so elegantly stated by John Hyde, an ex-Army aviator, Kit Fox owner, and retired Oakland FSS controller: When obtaining a briefing from an FSS or other source keep in mind that they're in "sales not production." In other words, don't blame the messenger for the message.

The following case study is from a NASA Aviation Safety Reporting system report.

CASE STUDY

I had departed Gallup, New Mexico, for Scottsdale, Arizona. I proceeded southwest toward better weather conditions than the direct route I had filed for my VFR flight plan. As I turned to a southerly heading it became clear to me that continued VFR toward Scottsdale was improbable. I called center and picked up an IFR clearance into Sedona, Arizona. As I climbed to my assigned altitude the aircraft started to pick up ice. At this point, I elected to cancel IFR and descend back to a lower altitude. I could see down through broken clouds to good visual ground contact. I turned northeast toward Winslow, Arizona, to find that continued VFR flight was, again, improbable. I was left with lowering clouds, darkening skies, and fuel becoming a factor. I picked up a southerly heading and after several minutes contacted center for radar vector into Payson, Arizona. I was switched to a different controller, who was reluctant to help because of my low altitude and the limited radar coverage in that area. I finally insisted upon "suggested radar headings to find the airport." Using the intermittent radar returns the controller was able to give me headings to the airport.

The ASRS report indicated the pilot had considerable experience, including flying jet aircraft. The pilot commented that: "I have never had a more challenging flight than that evening up on the Mogollon Rim northeast of Phoenix."

Could the pilot, with vast experience, have been complacent? Certainly the pilot pushed on into, admittedly, improble VFR weather. The pilot did not know the airplane's position, having to insist that ATC provide radar vectors to an airport. The pilot commented: "The springtime weather may be OK in Phoenix and Gallup, but the weather up on the Rim may not. I do not think I broke any rule, but I'll never do this again."

Did the pilot obtain a weather briefing for the flight as required by regulations? The pilot, apparently, was not aware that weather conditions can change drastically, especially in mountainous regions. Did the pilot maintain situation awareness of the airplane's position? What would have been the result, if the pilot had been unable to obtain radar assistance? Finally, in the event of an accident, what would have been the NTSB's probable cause?

Preparation

In order to apply a weather briefing to a flight, we must have done our homework. What is the terrain like along the route? What are the minimum altitudes? Are there suitable alternates? What if Plan A does not pan out? It's incumbent on the pilot—for every briefing, but especially if hazardous weather exists—to study the terrain, routes, and possible alternates for the proposed flight. The objective is to have an out. If there are no outs, the flight is a definite no go!

In order to apply a weather briefing to a flight, we must consider the aircraft we're planning to fly. Is the aircraft certified for the type of operation planned? For example, is the aircraft certified for VFR or IFR, day or night; is the aircraft certified for flight in known icing; does the aircraft have storm avoidance equipment? A negative answer, depending on the type of flight planned and the weather, may be a strong no go indicator.

How about the pilot and passengers? Human factors are often overlooked and can take on additional significance, depending on the type of flight planned and the weather. Certainly a night operation presents additional aircraft and pilot requirements and risks. An IFR operation puts additional workload on the pilot. The weather may be benign or horrendous. Each weather hazard puts additional risks on the flight. These risks may be mostly nullified or significantly increased by the equipment, or lack thereof, on the aircraft.

Evaluating the Weather

Among the many pilots that I have had the pleasure of serving at the Oakland FSS was a particular local air taxi pilot. He routinely flew from Oakland, through the Sacramento Valley, to northern California. He provided all the necessary background information and always requested a standard briefing. Since it's not part of a standard briefing, at the conclusion he would always ask for the closest area of clear conditions. What an excellent idea! This was one of the most prepared pilots I know. Should he have engine, navigation, or electrical problems, he knew the closest location of clear weather. We should all add this question to our personnel briefing requirements.

Some say never say never. Here's the exception that proves the rule. Never let the briefer make the go–no go decision. The briefer is a resource, and some are better than others. This applies equally to optimistic and pessimistic briefers.

For example, let's take the "VFR flight is not recommended" statement. It leaves considerable leeway for the briefer; some use this statement more than others. The inclusion of this statement should not necessarily be interpreted as an automatic cancellation, nor its absence as a go-for-it day. Notice that VNR applies to sky condition and visibility only. Few understand the provisions of special VFR. Hazardous phenomena, such as turbulence, icing, winds, and thunderstorms, of themselves, do not warrant the issuance of this statement. It is important to remember that this is a recommendation. Why then such a statement? It's simple; every year pilots insist on killing themselves and their passengers at an alarming and relatively constant rate by flying into weather where they have no business being.

The "VFR not recommended" statement was instituted in 1974, presumably because the last person a pilot would talk to was usually the briefer. A logical, although alarming, result of this statement is the increasing number of pilots who, in the absence of VNR, ask, "Is VFR recommended?" So far, the answer remains that the decision as to whether the flight can be safely conducted rests solely with the pilot.

It's been my experience that VFR flight is possible between 50 to 60 percent of the time that this statement has been issued for flights that I planned VFR. Remember that this is based on my training and experience. This doesn't mean it should be ignored, but we must take a careful look at the "complete picture." We'll revisit this issue in our discussion of risk assessment and management.

Updating Weather en Route

Updating weather en route begins with a complete or standard preflight briefing. In fact, many pilots will follow weather trends for several days before a planned flight. In this way, they get a feeling for the weather. Without this background knowledge, we will not be able to fully apply updated weather en route. For example, en route we obtain information that our destination is below our minimums. The next question becomes: Why? Is it due to the delayed improvement of stratus and fog, the faster-than-forecast advance of a frontal system, or the approach of a hurricane. We need to know in order to develop a sound alternate plan.

However, a pilot's responsibility does not end with an understanding of weather and forecasts, and a complete preflight briefing. Because of the dynamic character of the atmosphere, data must be continually updated. Surprisingly, many pilots have not been taught, or learned, the importance of updating weather reports and forecasts en route. Failure to exercise this pilot-in-command prerogative, as we've seen, can have disastrous results.

Up-to-date and accurate information, and a knowledge of how to apply it, is the key to an intelligent inflight weather decision. There are four basic sources of information:

- The pilot

- Flight Service Stations (FSSs)

- Enroute Flight Advisory Service (Flight Watch)

- Air Traffic Control facilities—centers and towers

The pilot, through direct observation and analysis of aircraft instruments, is a key source. Often the pilot is the only means of evaluating inflight weather conditions, especially cloud types and winds and temperatures aloft. To successfully apply the information requires a sound knowledge of aviation weather phenomena and theory.

FAA Flight Service Stations are a primary source of weather information. All of the information available during the preflight weather briefing is accessible inflight through FSS communications outlets. This information includes current observations, pilot weather reports (PIREPs), radar reports, satellite imagery, and, often, update forecasts.

Selected FSSs provide a continuous broadcast of weather advisories and urgent PIREPs over selected radio navigation aids (VOR). This service is known as Hazardous Inflight Weather Advisory Service (HIWAS). When a weather advisory affects an area within 150 miles of a HIWAS outlet, an alert is broadcast once on all frequencies, except Flight Watch and emergency.

The objective and purpose of Flight Watch is to enhance aviation safety by providing en route aircraft with timely and meaningful weather advisories. This objective is met by providing complete and accurate information on weather as it exists along a route pertinent to a specific flight, provided in sufficient time to prevent unnecessary changes to a flight plan, but, when necessary, to permit the pilot to make a decision to terminate the flight or alter course before adverse conditions are encountered.

Flight Watch is not intended for flight plan services, position reports, and initial or outlook briefings, nor is it to be used for obtaining aeronautical information, such as NOTAMs, center or navigational frequencies, or single or random weather reports and forecasts. Pilots requesting services not within the scope of Flight Watch will be advised to contact an FSS.

Using all sources, Flight Watch provides en route flight advisories, which include any hazardous weather, presented as a narrative summary of existing flight conditions—real-time weather—along the proposed route of flight, tailored to the type of flight being conducted.

The purpose of Flight Watch is to provide meteorological information for that phase of flight that begins after climbout and ends with descent to land, therefore the specialist can concentrate on weather trends, forecast variance, and hazards. Flight Watch is specifically intended to update information previously received and serve as a focal point for system feedback in the form of PIREPs. ARTCC and tower controllers do accept PIREPs, but weather is a secondary duty and, unfortunately, PIREPs aren't always passed along; if at all possible, PIREPs should be reported directly to Flight Watch. The effectiveness of Flight Watch is to a large degree dependent on this two-way exchange of information.

Enroute Flight Advisory Service (EFAS), originally Enroute Weather Advisory Service (radio call Eeewaas, which no one could pronounce), began on the West Coast in 1972 originally as a 24-hour service; Flight Watch now normally operates from 6 A.M. until 10 P.M. local time. Flight Watch is not available at all altitudes in all areas. The service provides communications generally at and above 5000 ft AGL, although, in areas of low terrain and closer to communication outlets, service will be available at lower altitudes.

The system expanded in 1976 and a network of 44 Flight Watch Control Stations became operational in 1979. The common frequency, 122.0 MHz, immediately became congested, especially from aircraft at high altitudes. A discrete high-altitude frequency was assigned Flight Watch stations in the southwest in 1980, to help resolve the problem. With Flight Service Station consolidation, Flight Watch responsibility has been assigned the FSSs associated with the Air Route Traffic Control Centers (Oakland FSS—Oakland Center, Hawthorne FSS—Los Angeles Center, etc.). A discrete high-altitude frequency, for use at and above Flight Level 180, has been assigned each Flight Watch Control Station to cover the associated center's area.

Establishing communications is the first step. Because only one frequency is available for low altitudes, pilots must exercise frequency discipline. In addition to the basic communications technique the following procedures should be used when contacting Flight Watch.

1. When known, use the name of the associate Air Route Traffic Control Center (ARTCC) followed by "Flight Watch" (Salt Lake Flight Watch). If not known, simply calling Flight Watch is sufficient.

2. State the aircraft position in relation to a major topographical feature or navigation aid (in the vicinity of Fresno, over the Clovis VOR, etc.).

    Exact positions are not necessary, but the general aircraft location is needed. Flight Watch facilities cover the same geographical areas as the ARTCCs. With numerous outlets, on a single frequency, the specialist needs to know which transmitter serves the pilot's area. This will eliminate interference with aircraft calling other facilities, garbled communications, and repeated transmissions. Failure to state the aircraft position on initial contact is the single biggest complaint from Flight Watch specialists.

3. When requesting weather or an en route flight advisory, provide the controller with cruising altitude, route, destination, and IFR capability, if appropriate.

The controller needs sufficient background information to provide the service requested.

Flight Watch controllers are required to continually solicit reports of turbulence, icing, temperature, wind shear, and upper winds regardless of weather conditions. This information, along with PIREPs of other phenomena, is immediately relayed to other pilots, briefers, and forecasters. Together with all sources of information the controller has access to the most complete weather picture possible.

ARTCC controllers are helpful in relaying reports of turbulence and icing, and providing advice on the location of convective activity, but the information is limited by equipment, and usually to immediate and surrounding sectors. Their primary responsibility is the separation of aircraft. On the other hand, Flight Watch has only one responsibility, weather. Flight Watch—with real-time National Weather Service weather radar displays, satellite pictures, and the latest weather and pilot reports—provides specific real-time conditions, as well as the big picture. Additionally, Flight Watch controllers have direct communications with Center Weather Service Unit personnel and NWS aviation forecasters.

Getting hold of Flight Watch is usually a simple matter, even for single-pilot IFR operations. ATC will almost always approve a request to leave the frequency for a few minutes, but don't wait until the last minute. Trying to find an alternate airport in congested approach airspace is no fun for anyone. I routinely use this procedure and have never been denied the request from en route controllers.

The early days of airline flying were plagued by thunderstorms as well as icing, turbulence, widespread low ceilings and visibilities, and the limited range of the aircraft. Today's jets have virtually overcome these obstacles. More and more pilots of general aviation aircraft, equipped with turbochargers and oxygen or pressurization, are encountering the same problems as yesterday's airline captains. The only difference is a vastly improved air traffic and communication system. Among one of the FAA's best kept secrets is the implementation of high-altitude Flight Watch.

Continually updating the weather picture is the key to managing a flight, especially at high altitude in aircraft without ice protection and storm avoidance equipment, and with relatively limited range. Winds aloft can be a welcome friend eastbound or a terrible foe westbound. With limited range, even a small change in winds at altitude can have a disastrous result. At the first sign of unexpected winds, Flight Watch should be consulted, if for no other reason than to provide a pilot report. A significant change in wind direction or speed is often the first sign of a forecast gone sour. A revised flight plan might be required. Flight Watch can provide needed additional information on current weather, PIREPs, and updated forecasts upon which to base an intelligent decision.

A primary reason for high-altitude flying is to avoid mechanical, frontal, and mountain wave turbulence; however, the flight levels have their own problems—wind shear and clear air turbulence. When problems are encountered, Flight Watch can help find a smooth altitude or alternate route. If the pilot elects to change altitude, an update of actual or forecast winds aloft is often a necessity.

Icing is normally not a significant factor in the flight levels, except around convective activity or in the summer when temperatures can range between 0°C and −10°C. However, icing can be significant during descent, especially when destination temperature is at or below freezing. Flight Watch can provide information on tops, temperatures aloft, reported and forecast icing, and current surface conditions.

Many aircraft are equipped with airborne weather radar and lightning detection equipment. However, these systems are plagued by low power, attenuation, and limited range. A pilot might pick his or her way through a convective area only to find additional activity beyond. Flight Watch has the latest NWS weather radar information. Well before engaging any convective activity, a pilot should consult Flight Watch to determine the extent of the system, its movement, intensity, and intensity trend. Armed with this information, the pilot can determine whether to attempt to penetrate the system or select a suitable alternate. ATC prefers issuing alternate clearances compared to handling emergencies in congested airspace and severe weather.

Finally, there is destination and alternate weather. The preflight briefing provided current and forecast conditions at the time of the briefing. This information should be routinely updated en route; the airlines do it, often through Flight Watch. Are updated reports consistent with the forecast? If not, why? Flight Watch controllers through their training are in an excellent position to detect forecast variance. Whether the forecast was incorrect or conditions are changing faster or slower than forecast, the pilot needs to know and plan accordingly. A knowledge of forecast issuance times is often helpful. Forecasts might not be amended if the next issuance time is close. Flight Watch is in the best position to provide the latest information and suggest possible alternatives.

Updates must be obtained far enough in advance to be acted upon effectively. This must be done before critical weather is encountered or fuel runs low. Hoping a stronger-than-forecast head wind will abate, or arriving over a destination that has not improved as forecast, is folly. At the first sign of unforecast conditions, Flight Watch should be consulted and, if necessary, an alternate plan developed. This might mean an additional routine landing en route, which is eminently preferable to, at best, a terrifying flight or, at worst, an aircraft accident.

High-altitude Flight Watch frequencies for individual ARTCCs are provided in Fig. 7-2. Frequencies and outlets also can be found on the

Fig. 7-2. Discrete high-altitude Flight Watch frequencies have been commissioned nationwide to eliminate much frequency congestion, especially with aircraft at low altitudes.

inside back cover of the *Airport/Facility Directory.* The standard frequency of 122.0 MHz can be used when a pilot is unsure of the discrete frequency.

The following case study is from a NASA Aviation Safety Reporting System report.

CASE STUDY

I checked the weather before I left Palo Alto, California, and the Central Valley was a mess with haze, but southern California was forecast to be fine for VFR flight into Fullerton. I followed I-5 from the Grapevine to Castaic. At this time it was obvious that the entire Los Angeles basin was covered by a layer at 3100 ft. I told the controller that I would like to follow I-5 to Fullerton. The controller said that may be possible at 2000 ft. I said fine and responded that I saw a hole about 10 miles ahead. When I dropped down, it was low visibility, but not technically IFR (3 to 5 miles haze). Having no desire to fly in these conditions, I climbed back to 7500 ft. I headed northeast to Lancaster and landed.

As in so many of the case studies, there is no evidence of the pilot checking weather with Flight Service or Flight Watch en route. The reporter commented: "When I saw the cloud layer my first thought was to return to Bakersfield." This whole incident would have been avoided if the pilot had gone with that thought. Countless accidents and incidents could be avoided if the pilot would turn around when that "thought" first occurs. Unfortunately, it's human nature to push on—get home-itis. This behavior, however, can be changed, but only by a conscientious effort on the part of the pilot. Like any addiction, the first step to a cure is admitting that there is a problem. These aren't called "sucker holes" for nothing. This pilot was fortunate to be able to file an ASRS report, rather than an accident report!

> **CASE STUDY**
> The pilot reported that a "grossly inaccurate weather forecast caused a planned VFR arrival at an unfamiliar airport to be made when surface-based Class E airspace was IFR due to low ceilings." The pilot reported being confused about the proper controlling agency for the airspace and that UNICOM was unresponsive. The pilot admitted to "insufficient planning for an alternate landing outside of controlled airspace."

It's difficult to assess what the pilot means by a "grossly inaccurate weather forecast." Certainly the pilot failed to update weather en route. The pilot was apparently untrained in determining the controlling agency for surface-based Class E airspace, and the scope and purpose of UNICOM.

By regulation, pilots are required to be trained in the "procurement and analysis or aeronautical weather reports and forecasts, including recognition of critical weather situations and estimating visibility while in flight." The pilot was unaware that the "controlling agency" can be found in the *Airport/Facility Directory* or by calling the Flight Service Station; or that UNICOM is a nongovernment communication facility which may provide airport information at certain airports. The pilot admitted not planning for an alternate. The pilot, presumably, came from an area of VFR weather, but blindly (pardon the pun) proceeded on.

Weather System Analysis

Now let's take a look at an actual weather event and apply the principles of radar and satellite interpretation. We'll take a look at February 14, 1998. The weather products used were observed between 1200Z and 1815Z, or generally early morning for the pacific and mountain regions, and late morning to midday for the central and eastern time zones. We'll use the Surface Analysis, Weather Depiction, and 500-mb (millibar) Constant-Pressure Charts, along with radar and satellite imagery. Since these products do not constitute a complete or standard briefing, this scenario cannot be used for a preflight weather decision. However, the analysis does illustrate the use of radar and satellite products as part of the "complete picture."

The Surface Analysis Chart provides a first look at weather systems. Constant-Pressure Charts analyze various upper layers of the atmosphere. Those usually used for aviation purposes depict layers at approximately 5000, 10,000, 18,000, 30,000, and 39,000 ft. The Weather Depiction Chart analyzes surface weather into three major weather categories: IFR, MVFR (marginal VFR), and VFR. Weather categories are contained in Table 7-1. The chart portrays general surface weather in a graphic form. The Radar Summary Chart depicts a summary of NEXRAD radar data.

These charts are available at Flight Service Stations and through most commercial vendors with graphics capability, although the actual data displayed may vary. For example, most commercial surface charts do not contain station model data. Charts are also available from the National Weather Service on the Internet at www.weather.noaa.gov/fax/nwsfax.shtml. The charts used in this text were obtained from this source.

The synopsis can be obtained from the 1500Z Surface Analysis chart in Fig. 7-3 and the 1200Z 500-mb Constant-Pressure Chart in Fig. 7-4. (The 500-mb Constant-Pressure Chart shows the flow patterns at approximately 18,000 ft, about halfway up through the atmosphere.)

Table 7-1. Weather Categories

CATEGORY	CEILING, FT		VISIBILITY, MILES
IFR	less than 1000	and/or	less than 3
MVFR	1000 to 3000	and/or	3 to 5
VFR	more than 3000	and	more than 5

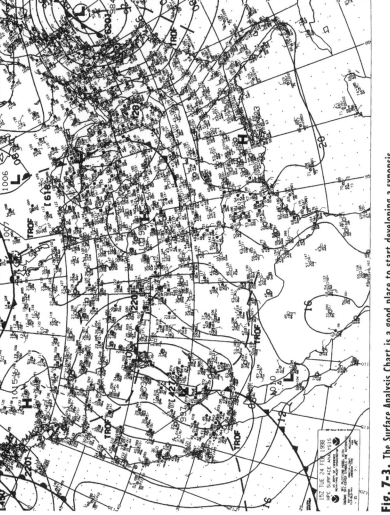

Fig. 7-3. The Surface Analysis Chart is a good place to start developing a synopsis.

Figure 7-3 shows surface winds blowing into the low in southern Nevada. Divergence—downward vertical motion—stabilizes the atmosphere, which decreases relative humidity and clouds. Figure 7-3 shows surface winds blowing out of the high in Alabama.

Note the anticyclonic, outward flow from the high centers over the midwest and Gulf coast in Fig. 7-3. A high or ridge implies surface divergence. These indicate areas of downward, stabilizing vertical motion in the lower part of the troposphere. Troughs, with their cyclonic flow, produce upward vertical motion. In Fig. 7-3, this is occurring along the trough in southern California.

A frontal system, an upward vertical motion producer, exists from the low pressure center over Great Salt Lake to the low center in southern Nevada, western Arizona, and offshore into the Pacific; a warm front associated with the low over Great Salt Lake extends through southern Wyoming, into northeastern Colorado. Another front extends from the Great Lakes into Indiana, southern Illinois and Missouri. Surface low pressure exists off the New England coast, with an associated occluded front and trough. These are areas of upward, destabilizing vertical motion in the lower part of the troposphere.

We would certainly expect an icing potential associated with the fronts over the Great Lakes and New England, with surface temperature at or above freezing. However, the clear skies associated with the cold front in southern Indiana, Illinois, and Missouri present little, if any, icing potential. The cold front associated with surface lows that run through Utah, Nevada, and western Arizona would be another area for potential icing, because of the low temperatures for this season.

In Fig. 7-3 upslope is occurring over the Plains in eastern Colorado and Wyoming, associated with the low over Salt Lake City and the warm front. Temperatures ahead of the warm front are below freezing. From this information alone, a pilot could expect to pick up a significant amount of ice over Wyoming that would be carried all the way to the ground. Another area of upslope is occurring in eastern Texas with the anticyclonic flow from the high along the Gulf coast.

An icing potential also exists in the New England region with overrunning warm air from the Atlantic Ocean over the colder land areas.

Pressure gradients for this day are weak over most of the 48 contiguous states, except in the northeast. Therefore, a pilot would not expect much in the way of mechanical turbulence on this day, except in the New England area.

Probably the most important and useful chart—maybe even more important to meteorologists than the surface analysis—the 500-mb chart describes the atmosphere in the middle troposphere, which is at an altitude of approximately 18,000 ft. This chart provides important pressure, wind flow, temperature, and moisture patterns, and can be used to determine areas of vertical motion at this level.

Unlike the surface analysis chart where upward vertical motion takes place along a trough line, upward vertical motion at the 500-mb level takes place between the trough and ridge line. Troughs transport cold air down from the north and warm air up from the south. Warm air rides northward on the east side of the trough—trough-to-ridge flow—so the air is lifted as it moves northward, producing upward vertical motion. When moisture is present clouds and precipitation in the midtroposphere develop. Conversely, in the ridge-to-trough flow, cold air sinks southward, producing downward vertical motion with typically clear, dry conditions in the midtroposphere. Clouds and precipitation frequently accompany upper-level lows and troughs, even without surface frontal or storm systems.

Moisture at the 500-mb level can be determined by darkened station models. This chart is a good indicator of high-level icing in summer months and of storms that are either well-developed or contain tropical moisture.

The 500-mb chart in Fig. 7-4 shows a trough along the west coast states that flows into a high-pressure ridge just east of the Rockies. The surface low pressure off New England is reflected as a closed low at 500 mb. This area of low pressure is almost vertical through the atmosphere, thus indicating a strong weather system. The 500-mb chart also shows shaded station model symbols, which indicate abundant moisture at this level. In fact, this is a classic Nor'easter. The trough over the west coast also shows abundant moisture in the trough-to-ridge flow, but the low is not cut off, indicating a weaker weather system. The ridge-to-trough flow over the midwest to the

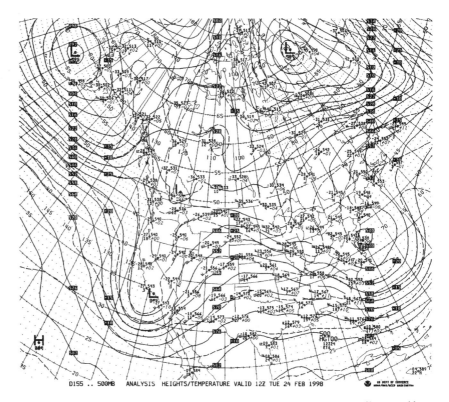

DISS .. 500MB   ANALYSIS HEIGHTS/TEMPERATURE VALID 12Z TUE 24 FEB 1998

Fig. 7-4. Along with the Surface Analysis Chart, the 500-mb Constant-Pressure Chart provides a look at weather systems halfway up through the atmosphere.

upper trough off the east coast is an area of downward, stabilizing vertical motion. This is also reflected in the open station models, indicating a dry air mass.

Maximum winds aloft occur from extreme southern California through Oklahoma and then southeast off the coast of northern Florida. This is the course of the jet stream. Since the jet is not located behind the surface frontal positions, either in the west or along the New England coast, these weather systems are not supported by the jet, and are thus weaker than if the jet was behind and supporting the vertical motion of these systems. Maximum clear air turbulence could be expected in the sharp trough over southern California, with the strongest winds associated with the jet stream, and with the sharp curvature of the contours off the east coast.

The Weather Depiction Chart is a computer-generated and -analyzed record of observed surface data. Frontal positions are obtained from the previous surface analysis. Information is hours old by the time the chart becomes available. The chart cannot consider terrain, nor is it intended to represent conditions between reporting locations. Gross errors between depicted categories and actual weather can occur. Since conditions could improve or deteriorate, data must always be updated with current reports. The Weather Depiction Chart is not a substitute for current observations.

Station models on the Weather Depiction Chart plot cloud height in hundreds of feet AGL, plotted beneath the model. Visibilities of 6 miles or less, and present weather, are entered to the left of the station. Sky cover and present weather symbols are the same as used on the surface analysis.

Refer to Fig. 7-5, a 1600Z, February 24, 1998, chart, produced 1 hour after the surface analyses in Fig. 7-3.

The Weather Depiction Chart provides a big, simplified picture of surface conditions. It alerts pilots to areas of potentially hazardous low ceilings and visibilities, and precipitation. The chart is often a good place to begin looking for an IFR alternate.

The warm front, discussed in the previous section, is weak. Conditions ahead of the front are typically VFR to MVFR, although snow is falling in western Wyoming. This would tend to discount any severe icing potential in this region. Weak upslope is also indicated. Cloud bases are VFR, lowering to MVFR, with some IFR conditions as the moisture flows toward the continental divide. The cold front in southern California is also weak, again tending to discount any serious weather hazards. However, serious weather problems can be anticipated in northern Arizona and Utah, associated with low ceilings and visibility and snow. Even in the areas depicted as MVFR, a VFR flight out of the valleys would be doubtful. Considerable rain and snow is associated with the systems over the Great Lakes and New England. This would certainly indicate serious icing potential in these areas, with widespread IFR conditions. Conditions over New Mexico and the southeast indicate little, if any, hazardous weather potential. Low ceilings and visibility appear to be a hazard to the VFR pilot in southeastern Texas.

The chart could be used to determine likely locations for a suitable alternate for an IFR flight into New England. An alternate to the

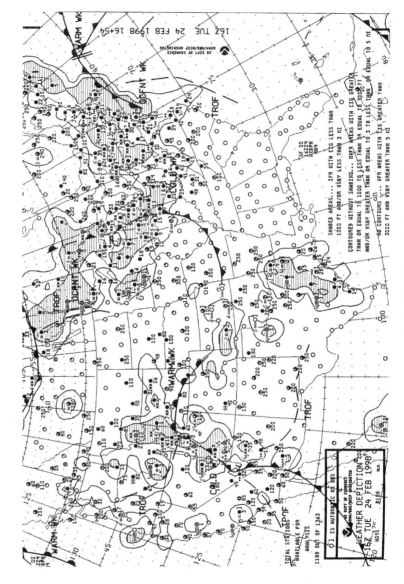

Fig. 7-5. The Weather Depiction Chart provides a big, simplified picture of surface conditions.

south, Maryland or Virginia, would be indicated. Michigan has good weather, but be careful; with a cold front to the west, conditions could change rapidly.

Refer to Fig. 7-6, a 1635Z, February 24, 1998, Radar Summary Chart. This time frame coincides with the Surface Analysis, Constant-Pressure, and Weather Depiction Charts previously discussed.

There is apparently considerable precipitation associated with the cold front in the southwest—apparently, because most of the precipitation is intensity level 1 and 2. Recall that this chart often depicts large, relatively scattered areas of weak or moderate precipitation as a solid band. Precipitation tops are also relatively low, with relatively high cloud bases, as indicated by the Weather Depiction Chart.

The weak warm front over Wyoming is producing very little precipitation; therefore serious icing, except in the area of light rain, is not likely. Here again, tops are quite low. The storm over the Great Lakes and New England is producing quite a bit of precipitation. Certainly, significant icing could be expected in this region. No thunderstorm activity was reported on this day, at this time. That's not unusual for the time of year. Precipitation tops in the New England region remain low.

No echoes are reported in eastern Texas, indicating shallow stratus and fog, but no precipitation. But note the NA symbol in south-central Texas. We'll have to use other sources to determine if any precipitation is occurring in this area.

With the preceding background let's move on to an interpretation of the satellite imagery for this day. Figures 7-7 and 7-8 contain four satellite images. The two in Fig. 7-7 are infrared and visible images from GOES-W (west). Those in Fig. 7-8 are infrared and visible images from GOES-E (east). They were taken on February 24, 1998 between 1815Z and 1830Z. They represent approximately the same general time period as the Surface Analysis Chart (Fig. 7-3), 500-mb Constant-Pressure Chart (Fig. 7-4), Weather Depiction Chart (Fig. 7-5), and Radar Summary Chart (Fig. 7-6).

Let's begin with Fig. 7-7. Its main feature is the frontal system depicted on the Surface Analysis Chart. It appears we have a rapidly

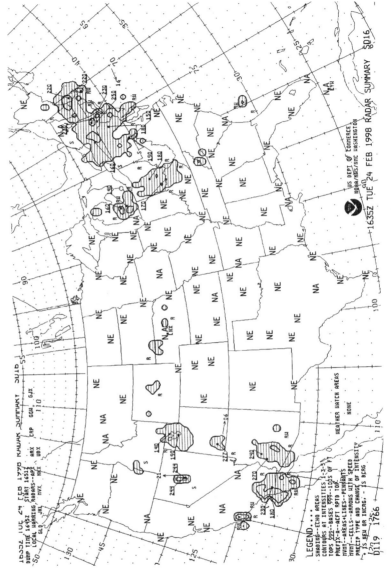

Fig. 7-6. The Radar Summary Chart often depicts large, relatively scattered areas of weak to moderate precipitation as a solid area.

moving cold front, since the satellite shows the frontal boundary already moving through central Arizona. There is an extensive cloud shield associated with the low over northern Utah, extending through Idaho, Montana, and the Dakotas. Tops are highest, associated with the front, over Arizona—represented by the brightest portion of the image. Note what appears to be low cloud—relatively dark—over western Texas in the IR image (left in Fig. 7-7).

Behind the front in central and northern California, the visible image indicates thick, white clouds, but the IR image is gray. These are relatively low clouds caused by upslope wind on the western side of the Sierra Nevada Mountains and the northern Tehachapi Mountains. Over the Pacific Ocean is an area of open-celled stratocumulus, associated with a weak low pressure circulation off the southern-central California coast.

Compare the cloud cover over southern California with the Radar Summary Chart in Fig. 7-6. The satellite image confirms that the area of precipitation is more scattered than depicted on the radar chart.

Refer to the IR image in Fig. 7-7. The brightness of the cold front confirms cold, thus high, tops. However, note how the frontal clouds are markedly dark for that portion of the front over the Pacific Ocean. In this area tops are quite low and the front is weak and diffuse. Tops are much lower over western Nevada and northern and central California. High tops are also associated with the cloud shield. This is certainly consistent with the Surface Analysis, Weather Depiction, and Radar Summary Charts. It also confirms less coverage of precipitation in southern California than indicated by the Radar Summary Chart. The southeast United States is clear, consistent with the high pressure over the area. Now look at northern and western Texas. What appeared in the visible image to be low clouds are revealed to have very cold tops. In fact, this is a thin band of cirrus. This signature, typically, represents the northern edge of the subtropical jet stream, and the clouds are called jet stream cirrus.

Refer to the visible image in Fig. 7-8. The cloud patterns are very consistent with the information depicted on the Surface Analysis, Weather Depiction, and Radar Summary Charts. Note considerable upslope on the west slopes of the Appalachian Mountains in West

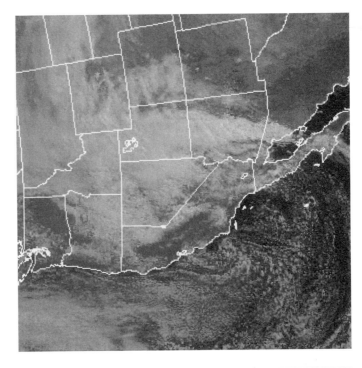

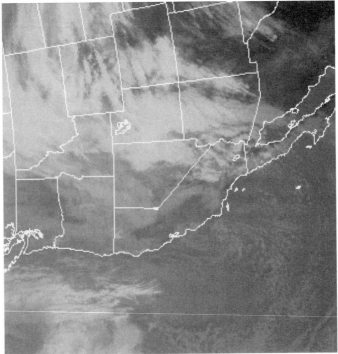

Fig. 7-7. These satellite images' main feature is the frontal system depicted on the Surface Analysis Chart.

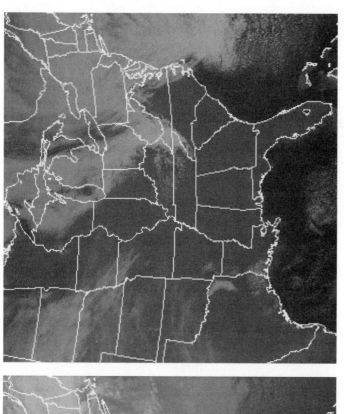

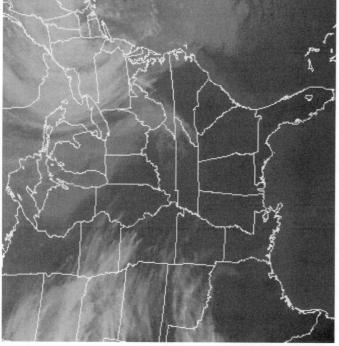

Fig. 7-8. The cloud patterns on these satellite images (left infrared, right visible) are very consistent with the information depicted on the Surface Analysis, Weather Depiction, and Radar Summary Charts.

Virginia, Virginia, and eastern Tennessee. Finally, note the cloud pattern in eastern Texas.

Now, refer to the IR image in Fig. 7-8. High tops are associated with the storm system over New England, relatively low tops over the Great Lakes and southern Appalachians. This is again consistent with Surface Analysis, Weather Depiction, and Radar Summary Charts.

With the surface winds obtained from the Surface Analysis Chart and information from the Radar Summary Chart, we can see some lake effect over the Great Lakes. I hope it can be seen how the application of all of these weather products is used to formulate this conclusion.

These satellite images are also consistent with the 500-mb Constant-Pressure Chart (Fig. 7-4). Clouds and precipitation in the trough-to-ridge flow over the western United States, clearing conditions in the ridge-to-trough flow over the midwest.

The satellite can often confirm or refute areas of weather depicted on various weather charts and predicted in forecast products. But, like every other product, its limitations must be understood, and the information must be taken along with all other available data. This is especially true at night when only the IR image is available.

Recall that the Weather Depiction Chart shows extensive areas of MVFR and IFR in southeastern Texas. The Surface Analysis Chart shows a southeastly flow at the surface along the Texas Gulf coast. The Gulf is an abundant source of moisture. From our knowledge of terrain, we know elevations increase toward the northwest. The satellite images confirm that these are low clouds—dark gray, warm tops on the IR image. These clouds are indeed low stratus and fog, caused by the anticyclonic flow around the high over the southeast and upslope. The visible image (Fig. 7-8) shows the area has almost dissipated along the Gulf coast, although clouds remain in northeast Texas. The stratus is not as extensive as depicted on the Weather Depiction Chart, a common anomaly of this product. There is certainly no icing or thunderstorm potential in this area. Since there is no significant cloud cover or vertical cloud development in south-central Texas, we can conclude that the NA on the Radar Summary Chart does not hide any hazardous weather.

CHAPTER 8

Evaluating Risk

A major theme over the last decade—to use a 1990s term—is risk management. Certainly the goal of every flight should be a safe outcome. (I often muse at controllers who ask: "How will this approach terminate?" I want to answer: "Successfully!")

At a recent FAA-sponsored accident prevention meeting our FSDO (Flight Standards District Office) Safety Program Manager asked the group to define *safe*. I responded, "risk free." Well, as we've seen the United States Supreme Court has determined that "safe" does not mean "risk free." In fact, safe can be defined as: "Not likely to cause or do harm or injury." What this means is that virtually every human endeavor involves some risk. As pilots, our goal is to reduce, as much as practical, our exposure to risk. In other words, how can we as pilots avoid becoming incident or accident statistics? We've only touched on this issue thus far; now we will focus the discussion on this subject.

We have reviewed the question of a pilot's training and experience, currency and the aircraft, weather and time of day, and physical and psychological condition. We have discussed the application of personal minimums, the interpretation and use of the weather briefing—with emphasis on radar and satellite products—and the necessity to update weather en route. Now we can move on to some practical methods of risk analysis and management—risk reduction.

How do we decide if a particular flight is safe? How do we assess the risk and manage that risk?

Accident prevention is part of the National Aeronautics and Space Administration's commitment to "aeronautics." To this end, NASA has developed scenarios of precursors to aviation accidents. A "precursor" is a factor that precedes, and indicates or suggests that an incident or accident will occur.

Refer to Fig. 8-1. Each "wheel" represents one precursor. It might be physical incapacity, poor judgment, aircraft deficiency, failure of the ATC system, the weather, or other factors which of themselves would not necessarily create an incident or accident, but when taken together have the potential to lead to disaster.

CASE STUDY

Seven-year-old Jessica Dubroff accompanied her father (a passenger) and the pilot in command (a flight instructor) in an attempt for a so-called transcontinental record involving 6660 miles of flying in eight consecutive days. (I say so-called record because this was nothing more than a publicity stunt. It reminds me of telling friends that my son soloed at age three months. He was the sole occupant of the airplane as we pulled it over to the wash rack.) The first leg of the trip, about 8 hours of flying, had been completed the previous day, which began and ended with considerable media attention.

On the second day they participated in media interviews, preflight, and then loaded the airplane. The pilot in command received a weather briefing which included weather advisories for icing, turbulence, and IFR conditions, due to a cold front moving through the area.

The airplane was taxied in rain for takeoff. While taxiing the pilot acknowledged receiving information that the wind was from 280° at 20 gusting to 30 knots. A departing Cessna 414 pilot reported moderate low-level wind shear, with airspeed excursions of ±15 knots. The airplane departed toward a nearby thunderstorm and began a gradual turn to an easterly heading.

Witnesses described the airplane's climb rate and speed as slow; they observed the airplane enter a roll and descent that was consistent with a stall. Density altitude at the airport was 6670 ft. The airplane's gross weight was calculated to be 84 pounds over the maximum gross weight at the time of impact.

The National Transportation Safety Board (NTSB) determined the probable cause was the pilot's improper decision to take off into deteriorating weather conditions. These included turbulence, gusty winds, an advancing thunderstorm, and possible carburetor and structural icing. The airplane was over gross weight. Density altitude was higher than the pilot was accustomed to. The result was a stall caused by the failure of the pilot to maintain adequate airspeed.

As in virtually all the previous examples, most incidents and accidents can be attributed to a series of relatively insignificant factors that, when taken together, cause the dilemma. Let's review the Dubroff accident in this context.

They were on a tight schedule. Publicity events had been scheduled in advance. The original takeoff time was delayed to allow Jessica additional sleep. The pilot was fatigued from the previous day's flight and obtained little rest during the night. The weather was marginal at best. The pilot had to obtain a special VFR clearance for departure. Who was really flying the airplane? The pilot in command was seated in the right seat of the Cessna Cardinal. Now add high-density altitude, an airplane over gross weight, and a mindset that they must go.

The first precursor was the need to keep a time schedule—sometimes referred to as "get-home-itis." Precursor number two was pilot fatigue. The next precursor was a high-density altitude takeoff with an airplane over gross weight. The fourth precursor was the weather, with its low ceilings and visibility, gusty winds, wind shear, turbulence, icing, and thunderstorms. (You could count each of these weather factors as an individual precursor.) A fifth precursor

NASA's ACCIDENT PRECURSOR SCENARIO

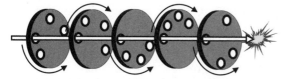

Alignment = Incident or Accident

Fig. 8-1. Precursors might be physical incapacity, poor judgment, aircraft deficiency, failure of the ATC system, or the weather.

was the pilot's attempt, under these very adverse conditions, to try to maintain control of the airplane. We will never know what exactly happened, but airplane control was lost. The deck was certainly stacked against them.

Like most accidents, I think we can see how breaking any one individual link could have prevented the mishap. The first link was the time schedule. A friend, and excellent pilot, has the philosophy that there is never a reason that you absolutely have to be anywhere. The Dubroff pilot's mindset appeared to be, "we're going no matter what."

The second link was fatigue. It was reported that Jessica had slept most of the first leg. As we've discussed, pilot fatigue is a significant factor in the deterioration of both mental and physical skills. This certainly may have clouded the pilot's go–no go decision, along with failure to calculate gross weight and density altitude.

The weather was terrible. If the weather had been clear and calm, the pilot might have gotten away with fatigue, overloading the airplane, and lack of experience with high-density altitude.

Now add the pressure of flying from the right seat, with a novice student in the left, in less than basic VFR conditions. Even a slight, momentary distraction under these conditions can have serious consequences. It's reasonable to conclude that the pilot experienced sensory overload during climbout. All of these factors together aligned the precursors, resulting in a fatal accident.

So how do we assess and manage risk? We apply aeronautical decision making—the ability to obtain all available, relevant information, evaluate alternative courses of action, then analyze and evaluate their risks, and determine the results. First, evaluate all the factors for a particular flight and decide if the risk is worth the mission. Our goal is to prevent the precursors from aligning. That's easy for me to say. This can be extremely simple or extremely complex. There are three elements in risk assessment and management. They are: planning, aircraft, and pilot. Planning is the "homework" part of the flight, as previously mentioned. We study terrain, altitude requirements, and the environment. The environment includes the weather, our personal minimums, and alternatives. Now we evaluate the aircraft. Does it have the performance and equipment for the mission? If the answer is yes, we preflight the aircraft and determine that it is airworthy for the mission—day and night, and VFR and IFR require different minimum aircraft equipment. Assuming the pilot is "fit for flight," we're ready to go. Simple, huh?

CASE STUDY

A pilot friend and I picked up a new Cessna 172 in Omaha, Nebraska. We departed about 10 A.M. for the return trip to Van Nuys, California. After takeoff we noticed the turn coordinator was not operating. We returned to Omaha, and about an hour later the instrument was replaced and we proceeded to our first destination—Garden City, Kansas.

After refueling the airplane and ourselves, we proceeded on to Albuquerque, New Mexico. By now it was late afternoon. I walked into the Albuquerque FSS and said to the controller, "Cancel a flight plan for an airplane ending in Quebec out of some place back east." It was a new airplane and I just couldn't remember where we had departed on this leg. The FSS controller invited us to stay over for the night.

Both myself and my partner were fatigued from our flight and it was about two hours before sunset. However, with two qualified pilots onboard, one of us could rest and "cat nap" while the other flew. We decided to fly toward Phoenix, Arizona rather than the normal route via Prescott. In this way we would be over flat terrain for that portion of the flight that would be conducted at night. Weather was not a factor and the remainder of the flight to Van Nuys was uneventful.

The preceding case study illustrates the aeronautical decision-making process. With two qualified and current pilots, the risk of fatigue was reduced. By flight planning over flat terrain, the risk of a night flight over the mountains was nullified.

A flight decision can be as simple as my friend John and his Kit Fox looking at an afternoon flight in the traffic pattern. (John Hyde is an ex-Army aviator, extremely competent pilot, recently retired from the Oakland FSS, and currently building a kit plane.) Or, as complex as one of NASA's Space Shuttle missions. To help us, I've developed the risk assessment and management decision tree, Fig. 8-2.

Let's start with John's decision. Planning: Airport elevation 397 ft, runway 25L 2699 ft; pattern altitude 1400 ft; the environment—clear,

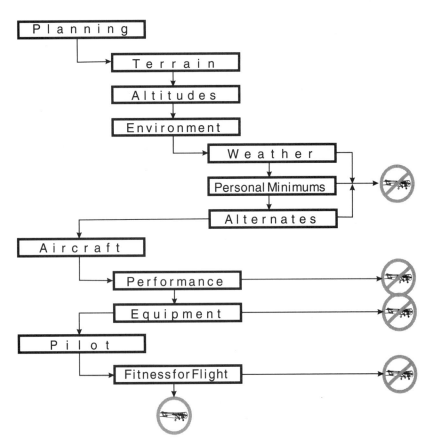

Fig. 8-2. Risk assessment and management decision tree.

cool, winds calm, alternate runway 25R. Aircraft: Performance of the Kit Fox OK; airplane equipped for day VFR flight in Class D airspace. Pilot: Fit for flight. Decision: Go!

Don't worry. We're not going to evaluate a Space Shuttle mission. Instead let's take an actual flight situation. We were flying from Oklahoma City to Palm Springs, California, for the 1998 AOPA convention. We made it as far as Santa Rosa, New Mexico, before the weather closed in. Hal Marx (USMC retired), the Santa Rosa airport manager, refueled our airplane and gave us a lift into town, where we remained overnight. The next day wasn't any better and we spent another night.

CASE STUDY

We had been trying to get to Albuquerque for 2 days without success. The following morning wasn't much better, but forecast to improve—how many times have we heard that?

Planning. Upslope wind due to rising terrain was, and continued to be, the culprit. Santa Rita has a field elevation of 4782 ft. Along I-40, the high plateau of eastern New Mexico rises to over 7000 ft, with the pass through the Sandia Mountains at about the same elevation. Terrain is slightly lower to the north and south, but still over 6000 ft. Because of the mountains IFR minimum en route altitudes (MEAs) vary from about 10,000 to 12,000 ft. Minimum altitudes for a VFR flight would range from 6500 to 8500 ft. We were flying a Cessna 172 that was equipped and certified for both day and night, and VFR and IFR.

Environment. MVFR to IFR ceilings, generally good visibility, high tops, freezing level at about 10,000 ft, conditions forecast to slowly improve during the day. The satellite image indicated widespread clouds, not uncommon for this type of weather event. Radar indicated only light to moderate precipitation, with no thunderstorms. In evaluating risk: Flying toward or in improving weather is better than flying toward or in deteriorating conditions.

With my training and experience I have different personal minimums depending on the environment. I also have confidence in my ability to make the decision to turn around. As my friend John puts it, "Cowardice is the better part of valor."

IFR flight. High minimum altitudes, low freezing level, I did not have approach charts for Albuquerque. The airplane would be at the limit of its performance envelope. The airplane was equipped for IFR operations, except not certified for flight in icing conditions. Thunderstorms not reported or forecast—stable air mass. Cloud bases at Albuquerque acceptable (MVFR) and forecast to improve. We would be at the MEA in probable icing conditions, unable to climb, over mountainous terrain—mountains obscured in clouds. Alternates should something go wrong—none! Risk high. Decision: No go.

VFR flight: Plan A—Climb to VFR on top and fly to Albuquerque and descend through broken clouds—forecast, anyway. Plan B—Fly under the clouds and land at Albuquerque. Plan C—Fly south, along the railroad to Albuquerque. (For some reason railroad engineers always select the lowest terrain.) Plan D—Return to Santa Rosa. Risk, yes; but plenty of options. For me this was a "go take a look" situation. Why? The area was sparsely populated and had good visibility, weather was good at the departure airport. On the negative side, I was not familiar with the area, although familiarity has led many a pilot to disaster. It was daylight. A night flight, either VFR or IFR, under these conditions would have resulted in a no go decision. Risks are increased flying over mountainous terrain at night in low-performance, single-engine aircraft.

Airplane performance and equipment was acceptable for the VFR plan. The pilot was fit for flight. Decision: Go.

Risk assessment and management does not stop with a go decision. We must re-evaluate conditions throughout the flight from preflight inspection to determining that a particular airport is suitable for landing. Should the airplane be unairworthy—this includes equipment—for the flight the decision would be no go. If conditions at the destination (wind, weather, surface conditions, etc.) change, we may have to divert. If we don't have an alternate plan risk is too high, resulting in a no go decision.

Part of the "complete picture" is having more than one way out. When only one out is left, it's exercised. This might mean canceling a flight, circumnavigating weather, avoiding hazardous terrain, or an additional landing en route. The 180° turn is made before entering clouds. If the situation becomes uncertain, assistance is obtained before an incident becomes an accident. In this way a good pilot will never be caught on top or run out of fuel. These pilots combine mental attitude and skill to update weather en route, devise a plan based on this information, and coordinate the action before the situation becomes critical.

CASE STUDY CONTINUED

With the preflight complete and 4¹/₂ hours of fuel, we departed Santa Rosa and opened our VFR flight plan to Albuquerque. (A VFR flight plan, especially under these conditions, is part of risk management.) Ceilings were low, but visibility was excellent. It soon became apparent that Plan A, over the clouds, was not going to work. This was confirmed by a conversation with Albuquerque Radio advising that their weather had not improved. Plan A: No go.

Plan B—Fly under the clouds. Approaching Clines Corners, terrain rises to about 7000 ft. The clouds went right down to the ground! When, as pilots, do we say no and call it a day? I teach, or maybe it's preach, that the first time the thought occurs "Should I really be here" or "Maybe I should turn around," it is a red flag to take positive action now! Don't push the weather, your aircraft, or yourself; turn around and wait it out. We initiated a 180° turn. We would have been flying from bad weather to worse weather. Risk too high. Decision: No go.

I had resigned myself to returning to Santa Rosa—Plan D. At this point my wife said, "What about plan C?" An increased risk accompanied Plan C. There were only a couple of dirt strips with high elevations and short runways for alternates. The terrain was lower, ceilings low, but visibility excellent. For navigation we had the "iron compass" (railroad). I called Albuquerque Radio and changed our route and estimated time of arrival (ETA). (If you're going to change your plans, it makes no sense not to update your flight plan with the FSS.) As is my practice, I made position reports and updated weather with flight service—another part of risk management. We always had the option of returning to Santa Rosa should the weather deteriorate. Albuquerque did not improve and we landed short at Alexander, New Mexico. With the weather now improving from the west, we continued on to Palm Springs.

The following case study is an example of using the system—FSS and ASOS, evaluating and managing risk, and the decision-making process. You may ask why I used flight service and not Flight Watch. Since flight plan revisions were required, Flight Watch was not appropriate. Also, I needed only specific weather reports. I was able to accomplish both tasks with one radio contact.

CASE STUDY

The Weather Channel reported a fast-moving cold front, with strong surface wind, approaching Palm Springs. Because of the large amount of traffic for the AOPA convention, there were no tie-downs available. (The airport people did, however, offer to sell us a set of commemorative chalks. Hum?) With our business concluded, we decided to leave on Saturday. I received a briefing and filed a VFR flight plan to Lancaster's Fox Field—Plan A.

It was a beautiful flight, smooth air with a high overcast. With plenty of fuel, we decided to continue on to Harris Ranch in the San Joaquin Valley. I updated weather and revised our flight plan with Riverside Radio—Plan B.

Over Bakersfield I made a position report to Rancho Radio and received updated weather. Navy Leemore, adjacent to Harris Ranch, was reporting visibility three-quarters of a mile in blowing dust, wind gusting to 35 knots! Well, we knew where the front was. We checked the weather at Visalia, and again amended our flight plan—Plan C.

As we proceeded toward Visalia we could see the billowing dust storm approaching from the northwest. Fortunately, Visalia had an ASOS. We were only 10 miles south of Visalia when the ASOS reported the arrival of the storm. I contacted Rancho Radio, advised them of the situation, cancelled the flight plan, and descended into the traffic pattern at Tulare—Plan D. We got the airplane tied down about 5 minutes before the storm hit.

At times it may be more efficient to plan shorter legs than normal when the weather is questionable. It's often extremely difficult to make a sound decision while flying the aircraft, with the pressure to continue the flight. It's always easier to evaluation the weather on the ground.

CASE STUDY

We were trying to get from Springfield, Illinois to Oklahoma City. Arriving in Saint Louis, we found that the ground-based weather radar showed cell after cell along our route to the southwest. We decided to wait it out until the next day. However, about 6 that evening the front passed and the weather cleared. We decided to continue on to Kansas City, which was also behind the front. As darkness fell, the southern horizon was ablaze with continual lightning.

Checking weather from Kansas City to Oklahoma City indicated clearing, but still thunderstorm activity. After departure we again saw lightning on the horizon. That night it just wasn't meant to be. We returned to Kansas City and spent the night. The next day was bright and clear. I was attending the FAA's Air Route Air Traffic Control School at the time. I missed half a day. It would have sure been more embarrassing to have been involved in an aircraft accident!

Continued VFR into adverse weather is an all-too-common probable cause. It is a sad commentary that all of these accidents are preventable. The following case studies are from NASA ASRS reports.

CASE STUDY

About 1 hour before the flight, I received a weather briefing for my route. The route was reported as VFR, except for some fog over Jackson, about 30 miles north of my course. I called back to file my flight plan. During climb I noted the city lights of my first checkpoint. It was about 1 hour before sunrise and the visibility was good. About 45 minutes into the flight, while looking down for my next checkpoint, I noticed that the ground lights began to flicker. When I looked up, the windscreen was gray. When I looked down, all the lights had vanished. I was in a cloud.

I had not expected to encounter this cloud, but there it was. I noticed some small patches of ice had formed on the windscreen. The visibility, or lack thereof, at this point seemed the least of my problems. I assessed that, according to training, I should climb to a higher altitude. That, as I recall, was preferred to descending, although each would probably work. If, however, I climbed and exited the cloud, I would still have no ground reference. I was thankful for a little over 3 hours of instrument training.

At this point I began a 180° turn, yet I remembered that I had flown through something that deposited ice, and did not want to re-encounter that event. What to do? I called Flight Service and reported that I had no visibility with ice on the windscreen. I also remember telling that I was only VFR certified. I asked for a report on the extent of this cloud, and which direction would be appropriate for me to exit and return to VFR conditions. They instructed me to descend to 2500 ft. I followed instruction and found that once again I could see the ground. I was next instructed to contact approach control. The flight continued uneventfully to destination.

This pilot was very fortunate to be able to file an ASRS report, rather than be involved in an accident that—according to statistics—would have been fatal. The pilot's decision to climb was not appropriate. It appears this was based on the pilot's interpretation that when icing is encountered climbing is the best response. This may be fitting for an IFR pilot, but in this case the proper procedure would be to immediately initiate a 180° turn.

> **CASE STUDY**
>
> I was flying at 600 to 700 ft AGL beneath ceilings of 1000 to 1500 ft when I encountered clouds. I suppressed a nearly irresistible desire to fly lower and get underneath. This would be clearly wrong; for all I knew this wasn't a cloud, but fog extending to the ground. I executed a standard rate turn for 1 minute, noted that my heading was 180° opposite the previous heading, and exercised patience while 2 or 3 unending minutes went by before returning to marginal Visual Meteorological Conditions (VMC). I landed safely, concluding the flight.

After inadvertently entering clouds the reporter made two sound decisions. First, do not descend. Since cloud bases were unknown this action could have very likely led to disaster. The second sound decision was to execute a standard rate 180° turn to the reciprocal heading, then continue on that heading. The reporter then describes "…2 or 3 unending minutes." In any anxious situation, every minute seems like 10! The reporter's third sound decision was to have patience while exiting the cloud.

The reporter went on to list the following contributing factors:

A desire to get home.

Darkening sky, making overall visibility worse.

Fatigue, from having flown all day.

Previous experience flying in marginal conditions.

An inappropriate level of optimism, given the situation.

This case study pretty much speaks for itself.

Thunderstorm Scenario

Figure 8-3 contains a weather radar image for 1825Z and a 1-km-resolution weather satellite image for 1800Z, observed on September 24, 2001.

This weather event is fairly typical for California during the late summer and early fall. Radar is showing a short, relatively thin line of thunderstorm extending from Sacramento in the north to east of San Jose in the south. Radar echoes are level 3 and level 4. The line has several breaks, which are indicated on the radar image. The large break in the north is

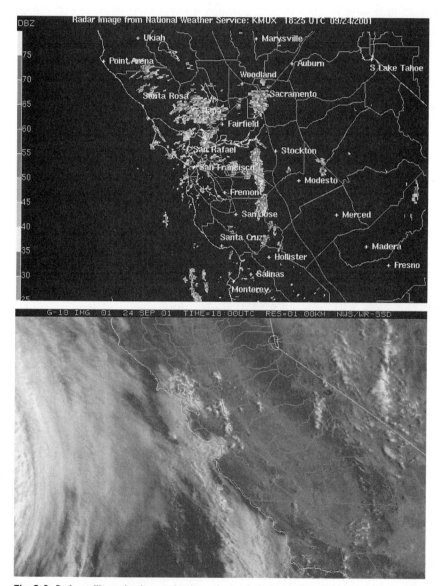

Fig. 8-3. Both satellite and radar are showing a short, relatively thin line of thunderstorm extending from Sacramento in the north to east of San Jose in the south.

cloud-free, as shown on the satellite image, but overcast with clouds east of San Jose. Another isolated area of thunderstorms is located between San Rafael and Fairfield. These echoes are also level 3 and level 4. Both areas are indeed thunderstorms, but neither area is severe.

The satellite image in Fig. 8-3 very closely correlates with the radar image. The shadowing and texture is clearly distinguishable from the shallow stratus along the coast and over the coastal waters. The distance between the isolated cells around Sacramento and the beginning of the line to the south is approximately 25 nm, as is the distance to the isolated cells northeast of San Rafael. The distance between the cells in the line east of San Jose is only about 10 nm.

The photograph in Fig. 8-4 is the scene as viewed from about 15 nm west of the line of thunderstorms. Cloud appearance is a very good

Fig. 8-4. Cloud appearance is a very good indicator of weather type and intensity.

indicator of weather type and intensity. Notice that the cloud tops are stratified cirrus, indicating the dissipating stage of a thunderstorm, as opposed to the cauliflower texture of towering cumulus—the cumulus and mature stage of a thunderstorm. These thunderstorms are rarely severe and limited state, which is typical of this area and weather pattern.

Now let's apply risk assessment and management to this weather event. In addition to the radar and satellite images in Fig. 8-3 and the visual picture of the clouds in Fig. 8-4, forecasts predict isolated nonsevere thunderstorms in this area. This confirms our analysis that the storms are not and will not become severe, covering relatively localized areas.

For flight out of the San Francisco Bay area to the east, we would want to avoid travel under the clouds and between the central portion of the line and the isolated thunderstorm east of San Jose. Therefore, we could proceed south to between San Jose and Hollister, then east. Since these cells are localized, visual avoidance and circumnavigation should not present a problem. A low-level flight east of Fairfield using visual avoidance would be acceptable. For high-performance airplanes, flight above these cells would be an acceptable plan. Conversely, flights approaching the Bay area can use the same rationale to avoid the hazardous weather.

CASE STUDY

We had planned a flight from Hayward, in the Bay area, to Ukiah. The time of year and weather were similar to that described in Fig. 8-3. We filed IFR in a Mooney 231 without storm detection equipment. By coordination with center we received approval for deviation around the thunderstorms, encountering only brief bouts with moderate turbulence.

A number of balloon pilots in the Napa Valley were not so fortunate. The gust fronts from the thunderstorms downed several of these fragile craft.

The radar and satellite images also tell us that the east side of the line is free of both precipitation and clouds. Pilots in flight can use

Flight Watch to obtain the same analysis en route, thus avoiding the hazardous weather with minimal rerouting or delay.

Lake Effect Scenario

The Chicago Area Forecast synopsis read:

> SYNOPSIS...11Z TROF YQT-SSM EWD. COLD CYCLONIC FLOW ACRS GRTLKS WITH LK EFFECT SHSN OVR UPR MI AND NRN LWR MI. LOW PRES OVR WRN WI WITH TROF NEWD TO NCNTRL ND. QSTNRY FNT RAP-IOW. 18Z QSTNRY FNT 80NW RAP-ONL THEN CDFNT ONL-IRK-JOT. 00Z HI PRES NERN ND. QSTNRY FNT RAP-OBH THEN CDFNT OBH-TTH. LK EFFECT SHSN CONTG OVR MI.

In part the synopsis decodes: "11Z...cold cyclonic flow across the Great Lakes with lake effect snow showers over upper Michigan and northern lower Michigan...00Z...lake effect snow showers continuing over Michigan."

Lake effect snow has developed along the downwind side of the Great Lakes on this late December day. This is a typical event for this time of year. Cold air flows over the still relatively warm waters of the lakes, resulting in snow and snow showers. Figure 8-5 contains infrared (left) and visible (right) satellite images of this event, observed at 1715Z.

From the satellite images we can see relatively low, thick cloud cover over eastern Lake Superior, upper Michigan, northern Lake Michigan and the northern portion of lower Michigan, and east into western New York. The next question becomes: How extensive are the snow showers? This information can only be inferred from the satellite images, but the Radar Summary Chart is an excellent "first look" for this data. Refer to Fig. 8-6, a portion of the Radar Summary Chart for this day, observed at 1715Z.

For the discussion, our main area of interest is lower Michigan and western New York. The Radar Summary Chart indicates an area of level 1 or level 2 echoes in northern Michigan with similar intensity echoes in western New York. The chart indicates the echoes are snow (S) and

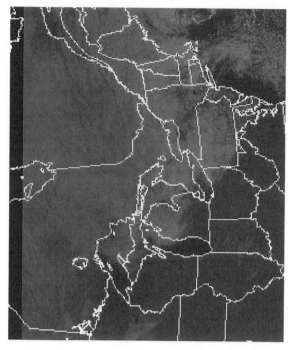

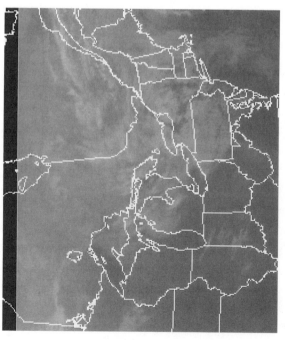

Fig. 8-5. The satellite image (left infrared, right visible) is a good place to start for an initial "picture" of cloud cover.

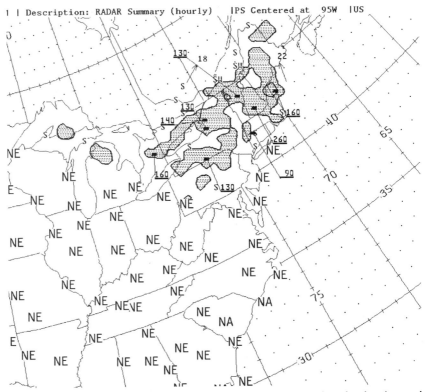

Fig. 8-6. Just as the satellite image is useful in determining cloud cover, the radar chart is a good place to start for an initial "picture" of precipitation coverage.

snow showers (SW). Maximum echo tops in western New York are reported to be 13,000 ft. Echoes are moving from the southwest at 18 knots. The chart depiction of areal coverage, intensity, and tops are what we would expect for this type of event. Note that the Radar Summary Chart indicates no echoes (NE) in both the Grand Rapids and Detroit areas.

For a more definitive radar picture we could consult some local radars, and apply our knowledge of the SD reports. Two local radar pictures are shown in Fig. 8-7. The left image is the Alpena County, Michigan site. On the right is an image from the Binghamton Regional, New York, radar. Both observations were taken at 1818Z.

From the scale in the left margin of the Alpena County image we see that the radar is in clear air mode. Therefore, even though the displayed intensity contains a great deal of yellow and red, they

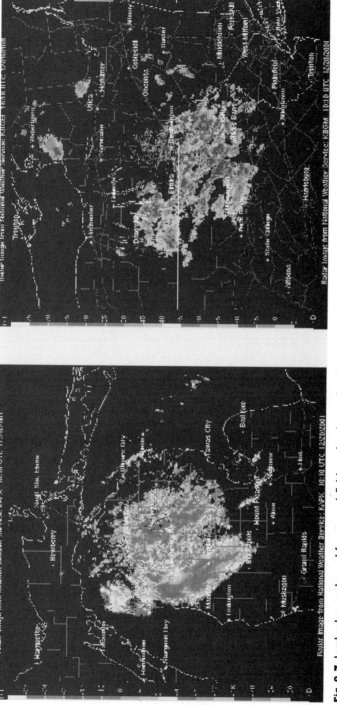

Fig. 8-7. Local radars can be used for a more definitive radar picture of precipitation.

259

represent only relatively weak echoes. (I know you can't see the color; you'll just have to take my word for it.) We can confirm the intensity with the Alpena County SD report below:

SDUS43 KWBC 201715
ROBAPX
APX 1735 AREA 6S 272/30 119/25 55W
AUTO
^LM2 MM22 NN1=

The SD reports an area with six-tenths coverage of moderate to light snow. (The intensity is obtained from the digital data at the bottom of the report.)

Even though the satellite indicates cloud cover farther south, toward Grand Rapids, both the Radar Summary Chart and Grand Rapids SD indicate the absence of any precipitation.

GRR—Kent County International Airport, Grand Rapids, Michigan
SDUS43 KWBC 201715
ROBGRR
GRR 1735 PPINE AUTO=

From the Binghamton, New York, image in Fig. 8-7 we see from the scale that the radar is in precipitation mode. The colors are blue and green. The intensity, again, can be verified from the Binghamton SD:

SDUS41 KWBC 201715
ROBBGM
BGM 1735 AREA 4S 279/95 127/66 102W MT 130 355/90
AREA 5S- 13/105 357/78 32W MT 130 355/90
AUTO
^IL1 JK11 LN211 MN121 NJ12111 OJ11 OM11 PK11 PN1=

In this case the digital data reveals that most of the snow is light intensity, with only some isolated areas of moderate snow.

Since lake effect snow typically contains large amounts of super-cooled water, significant icing is often a problem. We would have to consult AIRMET ZULU, SIGMETs, and appropriate Center Weather Advisories for specifics. The radar data indicates relatively low tops; this should be confirmed by PIREPs and the Area Forecast. Therefore, even airplanes with moderate performance should be able to top any icing hazards—but both climb and descent may present a significant icing problem.

Whiteout is an atmospheric optical phenomenon in which the pilot appears to be engulfed in a uniformly white glow. Neither shadows, horizon, or clouds are discernible; sense of depth and orientation is lost. Whiteout occurs over an unbroken snow cover and beneath a uniformly overcast sky, when light from the sky is about equal to that from the snow surface. Blowing snow may be an additional cause.

To the VFR pilot, whiteout is disastrous. Snow-covered terrain, an overcast, and already reduced visibilities are strong no go indicators. At the very first sign of this phenomenon the pilot's only option is a 180° turn to what, hopefully, will be better conditions.

IFR pilots are not immune from whiteout. For example, take the following accident.

The following is strictly speculation. Because the crash occurred after the second declaration of missing the approach, it appears the

CASE STUDY

The pilot's first destination did not have an instrument approach. It was snowing and the pilot reported "whiteout" conditions. The pilot diverted to an airport with an instrument approach.

Visibility was reported one-half variable between one-quarter and one mile in snow, indefinite ceiling 600 ft. Radio contact was lost after the pilot reported a second missed approach. Wreckage was located an hour later.

pilot decided to miss at the missed approach point. The transition between instrument and visual flight under these conditions is extremely difficult, especially with a single-pilot operation. The pilot may have acquired ground contact straight down, but slant range, and apparent whiteout conditions, would preclude visual contact with the approach environment and airport. Or, the approach environment and runway may have momentarily come into view, only to become obscured again in the variable visibility conditions.

Many commercial operations require two pilots just for this scenario. One pilot stays on the instruments and the other looks for visual cues. I had a similar experience while training an instrument pilot. We were flying at night in rain and fog; ceiling and visibility were at minimums. At the missed approach point my student started the missed approach. At that point I caught a glimpse of the approach lights. We were certainly not in a position to land and continued the missed approach procedure.

Personal minimums, and training and experience, play an important part in our flight decision. Certainly we would not want to place a student or low-time, inexperienced pilot in areas of even light snow. (Although it might be an opportunity for some takeoff and landing practice on a snow-covered runway, with a qualified instructor.)

An experienced pilot might elect to begin the flight with these conditions. Here is the dreaded: "Let's go take a look." There are no inherent additional risks in this operation as long as we continue to apply sound risk assessment and management techniques to the flight. In addition to those previously mentioned, consider flying a major highway, such as an interstate. After a snowfall, remember that the landscape will no longer look like the Sectional Chart. Many landmarks will most likely be covered with snow. Interstate highways are usually cleared of snow first, and as a last resort could be used as an emergency landing area. Following interstates does, however, usually eliminate a direct flight, although it reduces risk.

Do we have a current Sectional Chart? These are certainly not conditions to fly with a World Aeronautical Chart (WAC) or outdated Sectional. Carefully check for towers or other obstacles en route. Ensure there are suitable alternate landing fields. Here again, a nondirect route may reduce risk by providing additional suitable alternate landing areas. During the weather briefing, remember to check Notices to Airmen (NOTAMs) for all possible alternate landing airports.

En route risk assessment and management become strong players. Should we encounter snow, our only option may be an immediate retreat. We always should have an alternate in mind, should it become necessary. Don't attempt to penetrate the area of snow unless you can see the other side. Closely monitor the engine for possible carburetor or induction system ice. Check with Flight Watch or Flight Service en route for updates, and provide pilot reports.

The NTSB determined the probable cause to be the pilot's continued flight into adverse weather. They cited as factors the low

CASE STUDY

The Cessna 172 pilot received a preflight weather briefing that included marginal VFR conditions and reported icing in clouds near the route of flight. The pilot pulled the carburetor heat control to the on position, and descended to about 500 ft AGL to maintain visual contact with the ground. About $1/8$ inch of ice had formed on the airplane, and the pilot reversed course in an attempt to locate an airport. The pilot reported that the flight controls felt "sluggish." The pilot selected a field, configured the airplane, and made a precautionary landing. During the landing, the nose gear sank into the muddy field, then collapsed, and the airplane nosed over.

ceiling, icing conditions, airframe ice, and the muddy field—that is an alignment of at least four precursors. Unfortunately, this is an example of a situation where the pilot failed to retreat before entering adverse weather conditions. I know it's easy for me to say, but the goal is not to enter these conditions in the first place!

Let's consider the landing. Will we be forced to land on a snow- or ice-covered runway? This is certainly no place for a student or low-time, inexperienced pilot, or a pilot without adequate aeronautical charts. Do we have an alternate? We would not want to proceed without positive assurance of landing at our destination. This is easily accomplished with frequent updates en route from Flight Service or Flight Watch.

Upper-Level Weather System Scenario

Upper-level weather systems tend to modify and direct surface weather. They can intensify or stabilize conditions at the surface, cause thunderstorms to occur, and enhance or retard the intensity of frontal zones. Upper-level weather systems can cause severe conditions at the surface, or dampen or cancel out the vertical motion required to produce weather. The point is that not all weather is caused by frontal systems. In fact, nonfrontal weather-producing systems have considerable influence on surface conditions. A detailed explanation of these phenomena are contained in McGraw-Hill's companion publication, *Cockpit Weather Decisions.*

Often, nonfrontal weather-producing systems tend to confuse pilots and at times even weather briefers. This is because most weather texts tend to minimize or even ignore such phenomena. Thus far we've talked about upper-level troughs and upper-level lows, and the hurricane. Another example of a nonfrontal weather-producing system is the mesoscale convective complex (MCC). Mesoscale convective complexes are large, organized, homogeneous convective weather systems that can cover an area the size of several states. They tend to form during the morning hours.

Upper-level troughs and lows can affect any part of the contiguous United States. Hurricanes most often affect the Caribbean, Gulf coast, southern Atlantic coast, and at times the southwest United States. Mesoscale convective complexes most often affect the midwest states.

> **CASE STUDY**
>
> We had remained overnight in Las Vegas, Nevada, because of the weather. The next day was also blustery. The route to Van Nuys, California, was plagued with low clouds, mountain obscuration, and turbulence. I went into what was then the Las Vegas Flight Service Station for a briefing. After providing the briefer with necessary background information for the flight, this little old codger replied, "Well, you aren't going today!" This individual's technique was so obnoxious that without even thinking I replied, "Oh yes I am!" And, I hadn't even looked at the weather.

As is often the case in the winter, an upper-level low was over the area. There were no associated fronts, and this kind of system tends to bring poor weather for days. The briefer was correct in part; we were not going to fly a direct course to Van Nuys. Besides low ceilings and visibilities, the route was dominated with scattered rain showers and isolated thunderstorms. However, a careful check of the weather showed that the route from Las Vegas, south to Needles, California, then to Daggatt and Palmdale was feasible. This route has the lowest terrain, with plenty of alternate airfields. By circumnavigating the rain showers the flight was mostly smooth and without incident.

> ### CASE STUDY
>
> We had planned to attend the AOPA Convention in San Antonio, Texas. From Oxnard, California, we made it as far as El Paso, Texas. The weather between El Paso and San Antonio was dominated by an upper-level low, producing low ceiling and visibilities, icing, and embedded thunderstorms. Sometimes it's just not meant to be. We had to leave our Turbo-Cessna 150—that's an attempt at a little humor—and take one of American's 727s the rest of the way. Without ice protection and storm avoidance equipment, flight—VFR or IFR—under these conditions was out of the question.

The preceding incidents illustrate two upper-level weather systems. A careful check of the weather indicated one was a go and the other a no go. On the basis of the weather, the aircraft's equipment, and a complete weather briefing there really was no other rational decision possible. When the weather isn't flyable, don't go!

Figure 8-8 contains enhanced infrared and visible satellite images of a closed upper-level low off the Pacific coast. The remnants of the cold front are indicated by the cloud band extending along a north-south line from central Oregon, the northern California-Nevada border, and south to the vicinity of Los Angeles. The portion of the front in central and southern California is quite weak, with relatively low tops—the tops in this area have not reached the initial height for contouring on the enhanced IR image. One reason the front is weak, and continues to weaken, is a strong upper-level ridge just east of its position. The jet

Fig. 8-8. These satellite images illustrate a closed upper-level low off the Pacific coast. The remnants of the cold front are well east of the system.

stream is parallel to the front and is easily seen adjacent to the northern extent of cirrus clouds in the ridge's anticyclonic flow. The high cirrus band shows significant shadowing along its western edge in this midmorning visible image; the IR image shows well-defined contouring. The ridge axis is along a southwest Montana–Four Corners line.

The satellite image in Fig. 8-8 indicates that the center of the upper low is about 100 miles west of the northern California coast. Note the numerous cloud bands associated with the cyclonic circulation around the low center. On the east side of the low the clouds are thick with high, cold tops. These bands contain showers and isolated thunderstorms.

The location of the low can be verified by using the 500-mb Constant-Pressure Chart in Fig. 8-9. The center of the low is cut off—completely surrounded by a contour—and correlates very well with the satellite images. Note that the maximum winds at the 500-mb level

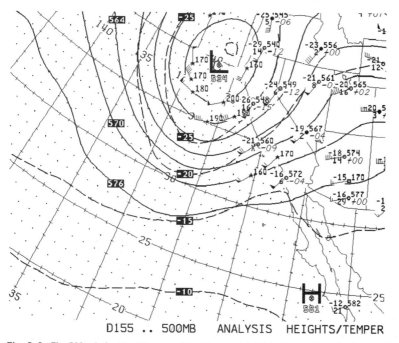

D155 .. 500MB   ANALYSIS   HEIGHTS/TEMPER

Fig. 8-9. The 500-mb Constant-Pressure Chart is very helpful in determining the flow midway through the atmosphere.

track around the low just north of Los Angeles, through Las Vegas, Nevada, then curve anticyclonically through northeastern Nevada and Idaho, then turn southeast through Montana and Wyoming. This verifies very well with the location of the jet stream and upper-level ridge as seen in the satellite images of Fig. 8-8. Notice over the Pacific Ocean that many of the station models use a star rather than a circle symbol. The stars indicate that the wind directions and speeds were determined using satellite imagery.

In addition to satellite images and charts, pilots have access to a description of the synopsis from the Area Forecast (FA) and, to a limited extent, Transcribed Weather Broadcast (TWEB) synopsis.

Below is the San Francisco FA synopsis for this weather event.

SYNOPSIS...STG UPR LOW 12Z MOV INTO WRN PTNS NRN CA CSTL WTRS FCST MOV ONSHORE NRN CA BY 00Z AND INTO NRN SIERNEV MTNS BY 06Z. AT SFC..PAC CDFNT 12Z NERN CA-SAN JOAQUIN VLY-SRN CA CSTL WTRS WL MOV EWD TO SRN NV-WRN AZ BY 00Z. SFC LOW 12Z NRN CA CSTL WTRS WL MOV EWD NR NRN CA CST BY 00Z. ..SMITH..

The synopsis states, in part, that there is a strong upper low at 1200Z that will move into the western portions of the northern California coastal waters and onshore northern California by 0000Z.

Below is the San Francisco TWEB forecast synopsis for the same period.

SFO SYNS 201402 LOW PRES SFC AND ALF WILL MOV INTO CNTRL CA BY 02Z. WNDS ALF MDT S-SW BECMG LGT SE-E NRN RTES BY 02Z AS TROF MOVES INTO CNTRL CA...CNTR OF LOW PRES SPRDG SHWRS AND ISOLD TSTMS OVR NRN AND CNTRL BY 21Z...AMS MOIST AND WKLY UNSTBL NR LOW.=

The TWEB synopsis, as is often the case, provides a more detailed explanation of the weather than is available from the FA synopsis. The TWEB synopsis indicates that the air mass is weakly unstable near the low center, which is expected to result in isolated thunderstorms.

The 1715Z Radar Summary Chart in Fig. 8-10 reveals an extensive area of light to moderate rain (R) and rain showers (RW) over central California, with some isolated thunderstorms (TRW). Precipitation tops range from 22,000 to 26,000 ft. Cell movement is generally from the south, reflecting the circulation around the low. Because of the relatively low detail available with this product, the chart, often,

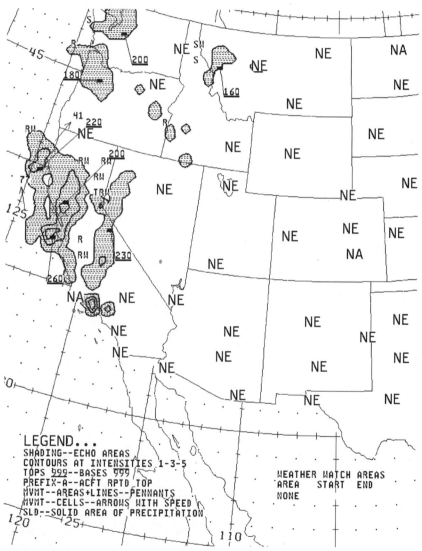

Fig. 8-10. Limitations of scale on the Radar Summary Chart prevent the detail available with an image from an individual radar site.

cannot show the detail that is available with an image from an individual radar site. However, it does provide a sense of the overall precipitation pattern—intensity, tops, and movement. Note that east of the ridge no precipitation is occurring. This is exactly what we would expect with the ridge's strong downward vertical motion.

The local 1755Z radar image in Fig. 8-11 provides a more detailed view of the precipitation within this weather system. Figure 8-11 shows several bands of weather. One band is onshore over the coastal sections, about to move into the San Joaquin Valley. Several other bands are approaching the coast. This is exactly what we would expect, from a knowledge of upper-level lows and the satellite images. The heaviest precipitation is occurring over the coastal mountains.

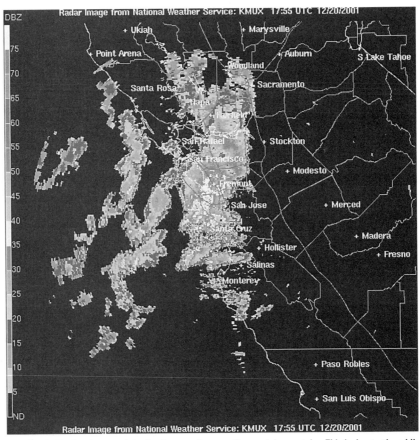

Fig. 8-11. The heaviest precipitation is occurring over the coastal mountains. This is due to the additional orographic upward vertical motion, which enhances the overall vertical motion of the system.

This is an area of additional orographic upward vertical motion, which enhances the overall vertical motion of the system, resulting in heavier showers and isolated thunderstorms.

With an analysis and understanding of the synopsis, let's take a look at an excerpt from the Area Forecast for central California.

CSTL SXNS...17-20Z BECMG SCT-BKN025 BKN050. TOPS FL200. OCNL VIS 3-5SM NMRS SHRA/ISOL TSRA. CB TOPS FL280. SAN JOAQUIN VLY...18-20Z BECMG BKN025 OVC050. TOPS FL200. OCNL VIS 3-5SM NMRS -SHRA/WDLY SCT TSRA. CB TOPS FL300. SIERNEV MTNS...OVC060-080 VIS 3-5SM -SN BR. OCNL SN. TOPS FL200.

Cloud bases in the FA are above mean sea level (MSL), unless noted as above ground level (AGL) or ceiling (CIG). All of the cloud bases in the above example are MSL heights. Also, remember that the FA is not a stand-alone product. To get the complete picture of the weather, the FA must be used along with the AIRMET Bulletin, specifically AIRMET SIERRA. On this day AIRMET SIERRA advertises occasional ceiling and visibility below 1000 ft and 3 miles, mountain occasionally obscured in clouds and precipitation.

For the coastal sections, the FA forecasts conditions to become between 1700Z and 2000Z: 2500 ft scattered to broken, 5000 ft broken, tops to 20,000 ft; visibility occasionally 3 to 5 miles in numerous rain showers and isolated thunderstorms, cumulonimbus tops to 28,000 ft. This verifies very well with the precipitation tops from the Radar Summary Chart.

This doesn't look too bad for a VFR flight over the coast sections. Or does it? A VFR pilot must still contend with the occasional IFR conditions and obscured mountains, as well as the showers and possible thunderstorms. So there is still considerable risk with a VFR operation in this area. How about an IFR flight? The low ceilings and visibility, and rain showers should not present much of a problem. But, again, don't forget about the thunderstorms and icing above the freezing level. This type of weather event is known to produce a serious icing hazard. However, with ice protection and thunderstorm avoidance equipment this weather system is not unduly risky, especially for multiengine high-performance airplanes.

The FA for the San Joaquin valley forecast similar conditions to the coast, except developing an hour or so later in the day with the approach of the low. VFR flight though the valley should be feasible, as long as the pilot can avoid the areas of rain showers and thunderstorms. This will require a careful weather watch, with an alternate airport close at hand. IFR operations will face the same concerns as those along the coast.

At quick look at the forecast for the Sierra Nevada mountains may not appear all that bad. Cloud layers are forecast to be between 6000 and 8000 ft. But remember these are MSL heights and the mountains and passes will be obscured in clouds and precipitation. VFR flight in the mountains is out. IFR flights will have to contend with the same conditions as the coastal sections and in the valley, except that the forecast does not expect thunderstorms to develop over the mountains. However, there would certainly be additional risk to single-engine or low-performance multiengine airplanes, should the engine fail.

Notice how the FA cannot, and does not, take into consideration the bands of weather that, from the satellite and radar products, we know will be moving through the area. Both VFR and IFR pilots must keep in mind that what may appear to be a break in the weather, or even the end of the storm, will only be followed with another significant band of weather. However, armed with a knowledge of radar and satellite interpretation, and a knowledge of forecast products, a pilot can apply sound, safe risk assessment and management techniques to his or her flying activity.

Requesting Assistance

All too often during incident and accident investigations, reporters mention their hesitation to declare an emergency. An emergency can be either a distress or urgency situation. Distress is a condition of being threatened by serious and/or imminent danger, requiring immediate assistance. Urgency is a condition of being concerned about safety and requiring timely, but not necessarily immediate assistance—a potential distress situation. Controllers would much prefer that pilots declare an emergency and obtain assistance before a bad situation becomes an impossible one.

CASE STUDY

It was the second day of the trip back from the Oshkosh Fly-In. We had departed Gillette, Wyoming, for Pocatello, Idaho, in a Bonanza. We planned to use VOR and pilotage navigation through Jackson Hole and Idaho Falls. The weather was not a significant factor, but visibility was restricted by smoke from numerous forest fires. In this part of the country, even at 12,500 ft we were still below the mountain peaks. After passing what I identified as the Grand Tetons we turned southwest toward Pocatello.

Well, you guessed it, we couldn't receive any VOR or establish communications with any facility. Continuing to fly down what I thought was the Snake River Valley, things didn't seem quite right. We were following an old aviation axiom: Follow a river or a road and it will normally bring you to a town, and hopefully an airport.

Even with $2^1/_2$ hours of fuel, we decided it was time to resolve the issue of position. Since we were unable to establish communications on standard frequencies, I selected 121.5. We were not in distress, but there was a sense of urgency. Therefore, as outlined in the *Aeronautical Information Manual* (AIM), I broadcast "PAN PAN PAN" followed by the aircraft identification. Almost immediately a military Air Evac flight responded. On the basis of our assumed position, the Snake River Valley, Air Evac provided us with a frequency for Salt Lake Center.

After several tries, however, we were unable to establish communication. By this time we had come across a small town with a good-size airport. Unfortunately, there was no name on the airport. Since our transponder was being interrogated, we knew someone had us on radar. I selected 7700 and again broadcast "PAN PAN PAN." Air Evac again responded. I requested Air Evac to ask center to look for a 7700 squawk. In a few moments Air Evac responded with another center frequency.

Calling center, the controller immediately responded, "Your position is 6 miles east of Big Piney." That left us with one minor question: Where is Big Piney? After a few moments shuffling the chart, we were on our way, although not by the route originally planned. As Maxwell Smart would say, "Missed it by that much!" Well, it isn't much on a WAC chart.

This incident illustrates several important points. En route, monitor the emergency frequency when you have a second radio. Keep careful track of your position and fuel. And, should a situation of uncertainty develop, don't hesitate to request assistance. The world doesn't come to an end by using 121.5 or squawking 7700. Unless there is some obvious pilot inadequacy, such as running out of fuel or flying into adverse weather, the FAA isn't going to get involved. Student pilots especially shouldn't fear requesting assistance. The FAA isn't going to talk to them, it's going to want to speak to their instructor. I know from personal experience. For my part, my students always know when and where to obtain assistance, as should every pilot. The goal is to prevent an incident from becoming an accident.

The FAA's policy: "Pilots who become apprehensive for their safety for any reason should *request assistance immediately.* Ready and willing help is available in the form of radio, radar, direction finding stations and other aircraft. Delay has caused accidents and cost lives. *Safety is not a luxury! Take action!*"

A popular aviation saying goes: "Aviation in itself is not inherently dangerous. But to an even greater degree than the sea, it is terribly unforgiving of any carelessness, incapacity, or neglect."

Weather Report and Forecast Contractions

Table A-1. Intensity, Proximity, and Descriptor

SYMBOL	MEANING	SYMBOL	MEANING
Intensity:		*Descriptor:*	
–	Light	TS	Thunderstorm
No symbol	Moderate	SH	Showers
+	Heavy	FZ	Freezing
Proximity:		DR	Low drifting
VC	Vicinity	BL	Blowing
		BC	Patches—*banc*
		MI	Shallow—*mince*
		PR	Partial

Table A-2. Precipitation

RA	Rain	GS	Small hail/snow pellets—*grésil*	SG	Snow grains
DZ	Drizzle	SN	Snow	IC	Ice crystals
GR	Hail (≥^{1}/4 in) —*grêle*	PL	Ice pellets	UP	Precipitation (automated observation)

Table A-3. Obstructions to Vision

FG	Fog (visibility $< {}^5/_8$ statute miles)	SA	Sand	PY	Spray
BR	Mist (visibility $\geq {}^5/_8$ statute miles) —*brume*	DS	Dust storm	SQ	Squall
HZ	Haze	SS	Sand storm	FC	Funnel cloud
FU	Smoke—*fumée*	VA	Volcanic ash	+FC	Tornado
DU	Dust	PO	Dust/sand whirls	+FC	Waterspout

Table A-4. Cloud Types

Ci	Cirrus	St	Stratus
Cs	Cirrostratus	Fra	Fractus
Cc	Cirrocumulus	Sc	Stratocumulus
As	Altostratus	Ns	Nimbostratus
Ac	Altocumulus	Cu	Cumulus
		Cb	Cumulonimbus

NWS Weather Radar Chart and Locations

The following list decodes NWS Weather Radar Chart locations shown in Fig. B-1.

Alabama

BMX	Birmingham/Alabaster WSR-88D/WFO
MOB	Mobile Regional Airport, Mobile WSR-88D/WFO

Arkansas

LZK	Little Rock WSR-88D/WFO
SRX	Ft. Smith WSR-88D

Arizona

EMX	Tucson WSR-88D
FSX	Flagstaff WSR-88D
IWA	Williams Gateway Airport, Phoenix WSR-88D
YUX	Yuma WSR-88D

California

BHX	Eureka WSR-88D
DAX	Davis WSR-88D
EYX	Edwards AFB WSR-88D
HNX	San Joaquin/Hanford WSR-88D/WFO
MUX	San Francisco WSR-88D
NKX	Miramar Naval Air Station, San Diego WSR-88D
SOX	Santa Ana Mountains WSR-88D
VTX	Los Angeles WSR-88D

NATIONAL WEATHER RADAR NETWORK

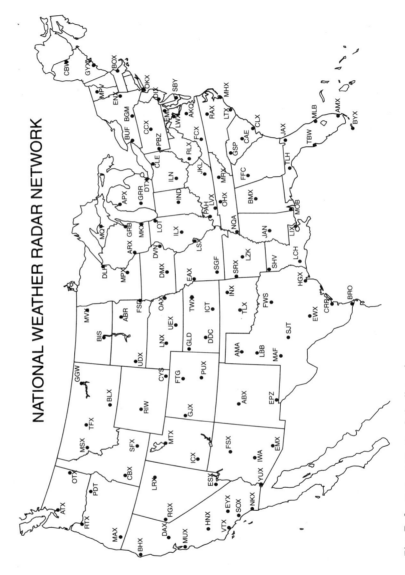

Fig. B-1. National Weather Radar Network.

Colorado
    FTG   Front Range Airport, Denver WSR-88D
    GJX   Grand Junction WSR-88D
    PUX   Pueblo WSR-88D
Florida
    AMX   Richmond Heights (Miami) WSR-88D/WFO
    BYX   Key West WSR-88D
    JAX   Jacksonville International Airport WSR-88D/WSO
    MLB   Melbourne International Airport WSR-88D/WSO
    TBW   Tampa Bay-Ruskin WSR-88D/WSO
    TLH   Tallahassee Regional Airport WSR-88D/WSO
Idaho
    CBX   Boise WSR-88D
    SFX   Pocatello WSR-88D
Georgia
    FFC   Peachtree City/Falcon Field Airport (Atlanta) WSR-88D/WSO
Illinois
    ILX   Central Illinois WSR-88D/WFO, Lincoln
    LOT   Lewis University Airport, Chicago/Romeoville WSR-88D/WFO
Indiana
    IND   Indianapolis International Airport, Indianapolis WSR-88D/WFO
    IWX   North Webster WSR-88D
Iowa
    DMX   Des Moines/Johnston WSR-88D/WFO
    DVN   Davenport Municipal Airport, Davenport WSR-88D/WFO
Kansas
    DDC   Dodge City Regional Airport, Dodge City WSR-88D/WFO
    GLD   Renner Field (Goodland Municipal) Airport, Goodland WSR-88D/WFO
    ICT   Wichita Mid-Continent Airport, Wichita WSR-88D/WFO
    TWX   Topeka WSR-88D

Kentucky
JKL	Julian Carroll Airport, Jackson WSR-88D/WFO
LVX	Fort Knox WSR-88D
PAH	Barkley Regional Airport, Paducah WSR-88D/WFO

Louisiana
LCH	Lake Charles Regional Airport, Lake Charles WSR-88D/WFO
LIX	New Orleans/Slidell WSR-88D/WFO
SHV	Shreveport Regional Airport, Shreveport WSR-88D/WFO

Maine
CBW	Houlton/Hodgton WSR-88D
GYX	Portland/Gray WSR-88D/WSO

Massachusetts
BOX	Taunton (Boston) WSR-88D/WSO

Michigan
APX	Green Township, Alpena County WSR-88D/WFO
DTX	Detroit/White Lake WSR-88D/WFO
GRR	Kent County International Airport, Grand Rapids WSR-88D/WFO
MQT	Marquette County Airport, Marquette WSR-88D/WFO

Minnesota
DLH	Duluth International Airport, Duluth WSR-88D/WFO
MPX	Minneapolis/Chanhassen WSR-88D/WFO

Mississippi
JAN	Jackson International Airport, Jackson WSR-88D/WFO

Missouri
EAX	Kansas City/Pleasant Hill WSR-88D/WFO
LSX	St. Louis/Weldon Spring WSR-88D/WFO
SGF	Springfield Regional Airport, Springfield WSR-88D/WFO

Montana
BLX	Billings WSR-88D
GGW	Glasgow International Airport, Glasgow WSR-88D/WFO

MSX	Missoula WSR-88D
TFX	Great Falls WSR-88D/WFO

Nebraska

LNX	North Platte WSR-88D
OAX	Omaha/Valley WSR-88D/WFO
UEX	Blue Hill (Hastings) WSR-88D/WFO

Nevada

ESX	Las Vegas WSR-88D
LRX	Elko WSR-88D
RGX	Reno WSR-88D

New Jersey

DIX	Wrightstown (Philadelphia) WSR-88D

New Mexico

ABX	Albuquerque WSR-88D
EPZ	Santa Teresa (El Paso) WSR-88D

New York

BGM	Binghamton WSR-88D/WSO
BUF	Buffalo WSR-88D/WSO
ENX	State University of New York/East Berne (Albany) WSR-88D/WSO
OKX	Brookhaven National Laboratory, New York City WSR-88D/WSO

North Carolina

LTX	Shallotte (Wilmington) WSR-88D
MHX	Morehead City/Newport WSR-88D/WSO
RAX	Raleigh/Durham WSR-88D

North Dakota

BIS	Bismarck Municipal Airport, Bismarck WSR-88D/WFO
MVX	Grand Forks WSR-88D

Ohio

CLE	Cleveland-Hopkins International Airport WSR-88D/WSO
ILN	Wilmington Airborne Airpark Airport WSR-88D/WSO

Oklahoma

INX	Inola (Tulsa) WSR-88D
TLX	Twin Lakes/Midwest City (Oklahoma City) WSR-88D

Oregon

MAX	Medford WSR-88D
PDT	Eastern Oregon Regional at Pendleton Airport, Pendleton WSR-88D/WFO
RTX	Portland WSR-88D

Pennsylvania

CCX	Centre County (State College) WSR-88D
PBZ	Corapolis (Pittsburgh) WSR-88D/WSO

South Carolina

CAE	Columbia WSR-88D/WSO
CLX	Charleston WSR-88D
GSP	Greenville/Spartenburg WSR-88D/WSO

South Dakota

ABR	Aberdeen Regional Airport, Aberdeen WSR-88D/WFO
FSD	Joe Foss Field Airport, Sioux Falls WSR-88D/WFO
UDX	Rapid City WSR-88D

Tennessee

MRX	Knoxville/Morristown WSR-88D/WFO
NQA	Millington Municipal Airport, Millington (Memphis) WSR-88D
OHX	Old Hickory (Nashville) WSR-88D/WFO

Texas

AMA	Amarillo International Airport, Amarillo WSR-88D/WFO
BRO	Brownsville/South Padre Island International Airport, Brownsville WSR-88D/WFO
CRP	Corpus Christi International Airport, Corpus Christi WSR-88D/WFO
EWX	San Antonio/New Braunfels WSR-88D/WFO
FWS	Fort Worth Spinks Airport, Fort Worth WSR-88D
HGX	Dickinson (Houston) WSR-88D/WFO
LBB	Lubbock International Airport, Lubbock WSR-88D/WFO
MAF	Midland International Airport, Midland WSR-88D/WFO
SJT	Mathis Field Airport, San Angelo WSR-88D/WFO

Utah
  ICX   Cedar City WSR-88D
  MTX   Salt Lake City WSR-88D
Virginia
  AKQ   Wakefield (Richmond) WSR-88D/WSO
  FCX   Roanoke WSR-88D
  LWX   Sterling (Washington D.C.) WSR-88D/WSO
Washington
  ATX   Everett WSR-88D
  OTX   Spokane WSR-88D
West Virginia
  RLX   Charleston WSR-88D/WSO
Wisconsin
  ARX   LaCrosse Ridge WSR-88D/WFO
  GRB   Austin Straubel International Airport, Green Bay WSR-
              88D/WFO
  MKX   Milwaukee/Sullivan Township WSR-88D/WFO
Wyoming
  CYS   Cheyenne Airport, Cheyenne WSR-88D/WFO
  RIW   Riverton Regional Airport, Riverton WSR-88D/WFO

Weather Advisory Plotting Chart and Locations

Figure C-1 is a Weather Advisory Plotting Chart. The following list decodes Weather Advisory Plotting Chart locations.

ABI	Abilene	TX
ABQ	Albuquerque	NM
ABR	Aberdeen	SD
ABY	Albany	GA
ACK	Nantucket	MA
ACT	Waco	TX
ADM	Ardmore	OK
AEX	Alexandria	LA
AIR	Bellaire	OH
AKO	Akron	CO
ALB	Albany	NY
ALS	Alamosa	CO
AMA	Amarillo	TX
AMG	Alma	GA
ANW	Ainsworth	NE
APE	Appleton	OH

WEATHER ADVISORY PLOTTING CHART

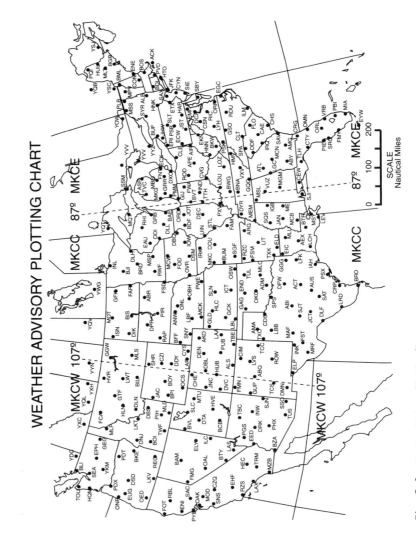

Fig. C-1. Weather Advisory Plotting Chart.

ARG	Walnut Ridge	AR
ASP	Oscoda	MI
ATL	Atlanta	GA
AUS	Austin	TX
BAE	Milwaukee	WI
BAM	Battle Mountain	NV
BCE	Bryce Canyon	UT
BDF	Bradford	IL
BDL	Windsor Locks	CT
BFF	Scottsbluff	NE
BGR	Bangor	ME
BIL	Billings	MT
BIS	Bismark	ND
BJI	Bemidji	MN
BKE	Baker	OR
BKW	Beckley	WV
BLI	Bellingham	WA
BNA	Nashville	TN
BOI	Boise	ID
BOS	Boston	MA
BOY	Boysen Reservation	WY
BPI	Big Piney	WY
BRD	Brainerd	MN
BRO	Brownsville	TX
BTR	Baton Rouge	LA
BTY	Beatty	NV
BUF	Buffalo	NY
BUM	Butler	MO
BVL	Booneville	UT

BVT	Lafayette	IN
BWG	Bowling Green	KY
BZA	Yuma	AZ
CAE	Columbia	SC
CDS	Childress	TX
CEW	Crestview	FL
CHE	Hayden	CP
CHS	Charleston	SC
CIM	Cimarron	NM
CLE	Cleveland	OH
CLT	Charlotte	NC
CON	Concord	NH
COU	Columbia	MO
CRG	Jacksonville	FL
CRP	Corpus Christi	TX
CSN	Cassanova	VA
CTY	Cross City	FL
CVG	Covington	KY
CYN	Coyle	NJ
CYS	Cheyenne	WY
CZI	Crazy Woman	WY
CZQ	Fresno	CA
DBL	Eagle	CO
DBQ	Dubuque	IA
DBS	Dubois	ID
DCA	Washington	DC
DDY	Casper	WY
DEC	Decatur	IL
DEN	Denver	CO

DFW	Dallas–Ft. Worth	TX
DIK	Dickinsin	ND
DLF	Laughlin AFB	TX
DLH	Duluth	MN
DLL	Dells	WI
DLN	Dillon	MT
DMN	Deming	NM
DNJ	McCall	ID
DPR	Dupree	SD
DRK	Prescott	AZ
DSD	Redmond	WA
DSM	Des Moines	IA
DTA	Delta	UT
DVC	Dove Creek	CO
DXO	Detroit	MI
DYR	Dyersburg	TN
EAU	Eau Claire	WI
ECG	Elizabeth City	NC
ECK	Peck	MI
EED	Needles	CA
EHF	Bakersfield	CA
EIC	Shreveport	LA
EKN	Elkins	WV
ELD	El Dorado	AR
ELP	El Paso	TX
ELY	Ely	NV
EMI	Westminster	MD
END	Vance AFB	OK
ENE	Kennebunk	ME

ENI	Ukiah	CA
EPH	Ephrata	WA
ERI	Erie	PA
ETX	East Texas	PA
EUG	Eugene	OR
EWC	Ellwood City	PA
EYW	Key West	FL
FAM	Farmington	MO
FAR	Fargo	ND
FCA	Kalispell	MT
FLO	Florence	SC
FMG	Reno	NV
FMN	Farmington	NM
FMY	Ft. Meyers	FL
FNT	Flint	MI
FOD	Ft. Dodge	IA
FOT	Fortuna	CA
FSD	Sioux Falls	SD
FSM	Ft. Smith	AR
FST	Ft. Stockton	TX
FWA	Ft. Wayne	IN
GAG	Gage	OK
GCK	Garden City	KS
GEG	Spokane	WA
GFK	Grand Forks	ND
GGG	Longview	TX
GGW	Glasgow	MT
GIJ	Niles	MI
GLD	Goodland	KS

GQO	Chattanooga	TN
GRB	Green Bay	WI
GRR	Grand Rapids	MI
GSO	Greensboro	NC
GTF	Great Falls	MT
HAR	Harrisburg	PA
HBU	Gunnison	CO
HEC	Hector	CA
HLC	Hill City	KS
HLN	Helena	MT
HMV	Holston Mountain	TN
HNK	Hancock	NY
HNN	Henderson	WV
HQM	Hoquiam	WA
HTO	East Hampton	NY
HUL	Houlton	ME
HVE	Hanksville	UT
HVR	Havre	MT
IAH	Houston International	TX
ICT	Wichita	KS
IGB	Bigbee	MS
ILC	Wilson Creek	NV
ILM	Wilmington	NC
IND	Indianapolis	IN
INK	Wink	TX
INL	International Falls	MN
INW	Winslow	AZ
IOW	Iowa City	IA
IRK	Kirksville	MO

IRQ	Colliers	SC
ISN	Williston	ND
JAC	Jackson	WY
JAN	Jackson	MS
JCT	Junction	TX
JFK	New York/Kennedy	NY
JHW	Jamestown	NY
JNC	Grand Junction	CO
JOT	Joliet	IL
JST	Johnstown	PA
LAA	Lamar	CO
LAR	Laramie	WY
LAS	Las Vegas	NV
LAX	Los Angeles International	CA
LBB	Lubbock International	TX
LBF	North Platte	NE
LBL	Liberal	KS
LCH	Lake Charles	LA
LEV	Grand Isle	LA
LFK	Lufkin	TX
LGC	La Grange	GA
LIT	Little Rock	AR
LKT	Salmon	ID
LKV	Lakeview	OR
LOU	Louisville	KY
LOZ	London	KY
LRD	Laredo	TX
LVS	Las Vegas	NV
LWT	Lewistown	MT

LYH	Lynchburg	VA
MAF	Midland	TX
MBS	Saginaw	MI
MCB	McComb	MS
MCK	McCook	NE
MCN	Macon	GA
MCW	Mason City	IA
MEI	Meridian	MS
MEM	Memphis	TN
MGM	Montgomery	AL
MIA	Miami	FL
MKC	Kansas City	MO
MKG	Muskegon	MI
MLC	McCalester	OK
MLD	Malad City	ID
MLP	Mullan Pass	ID
MLS	Miles City	MT
MLT	Millinocket	ME
MLU	Monroe	LA
MOD	Modesto	CA
MOT	Minot	ND
MPV	Montpelier	VT
MQT	Marquette	MI
MRF	Marfa	TX
MSL	Mussel Shoals	AL
MSP	Minneapolis	MN
MSS	Massena	NY
MSY	New Orleans	LA
MTU	Myton	UT

MZB	Mission Bay	CA
OAK	Oakland	CA
OAL	Coaldale	NV
OBH	Wolbach	NE
OCS	Rocksprings	WY
ODF	Toccoa	GA
ODI	Nodine	MN
OED	Medford	OR
OKC	Oklahoma City	OK
OMN	Ormond Beach	FL
ONL	O'Neil	NE
ONP	Newport	OR
ORD	O'Hare International	IL
ORF	Norfolk	VA
ORL	Orlando	FL
OSW	Oswego	KS
OVR	Omaha	NE
PBI	West Palm Beach	FL
PDT	Pendleton	OR
PDX	Portland	OR
PGS	Peach Springs	AZ
PHX	Phoenix	AZ
PIE	Saint Petersburg	FL
PIH	Pocatello	ID
PIR	Pierre	SD
PLB	Plattsburgh	NY
PMM	Pullman	MI
PQI	Presque Isle	ME
PSB	Phillipsburg	PA

PSK	Dublin	VA
PSX	Palacios	TX
PUB	Pueblo	CO
PVD	Providence	RI
PWE	Pawnee City	NE
PXV	Pocket City	IN
PYE	Point Reyes	CA
RAP	Rapid City	SD
RBL	Red Bluff	CA
RDU	Raleigh-Durham	NC
REO	Rome	OR
RHI	Rhinelander	WI
RIC	Richmond	VA
ROD	Rosewood	OH
ROW	Roswell	NM
RWF	Redwood Falls	MN
RZC	Razorback	AR
RZS	Santa Barbara	CA
SAC	Sacramento	CA
SAT	San Antonio	TX
SAV	Savannah	GA
SAX	Sparta	NJ
SBY	Salisbury	MD
SEA	Seattle	WA
SGF	Springfield	MO
SHR	Sheridan	WY
SIE	Sea Isle	NJ
SJI	Semmnes	AL
SJN	St. Johns	AZ

SJT	San Angelo	TX
SLC	Salt Lake City	UT
SLN	Salina	KS
SLT	Slate Run	PA
SNS	Salinas	CA
SNY	Sidney	NE
SPA	Spartanburg	SC
SPS	Wichita Falls	TX
SQS	Sidon	MS
SRQ	Sarasota	FL
SSM	Sault Ste. Marie	MI
SSO	San Simon	AZ
STL	St. Louis	MO
SYR	Syracuse	NY
TBC	Tuba City	AZ
TBE	Tobe	CO
TCC	Tucumcari	NM
TCS	Truth or Consequences	NM
TLH	Tallahassee	FL
TOU	Neah Bay	WA
TRM	Thermal	CA
TTH	Terre Haute	IN
TUL	Tulsa	OK
TUS	Tucson	AZ
TVC	Traverse City	MI
TWF	Twin Falls	ID
TXK	Texarkana	AR
TXO	Texico	TX
UIN	Quincy	IL

VRB	Vero Beach	FL
VUZ	Vulcan	AL
VXV	Knoxville	TN
YDC	Princeton	BC
YKM	Yakima	WA
YOW	Ottawa	ON
YQB	Quebec	QB
YQL	Lethbridge	AB
YQT	Thunder Bay	ON
YQV	Yorkton	SK
YSC	Sherbrooke	QB
YSJ	St. John	NB
YVV	Wiarton	ON
YWG	Winnipeg	MB
YXC	Cranbrook	BC
YXH	Medicine Hat	AB
YYN	Swift Current	SA
YYZ	Toronto	ON

Geographical Area Designators

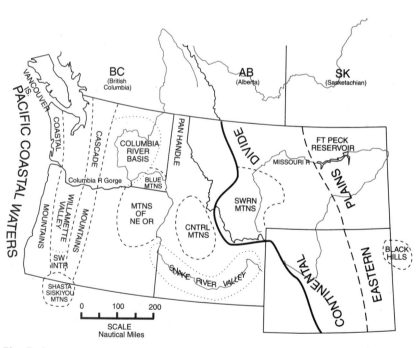

Fig. D-1. Common geographical area designators (northwest).

Fig. D-2. Common geographical area designators (southwest).

Fig. D-3. Common geographical area designators (central).

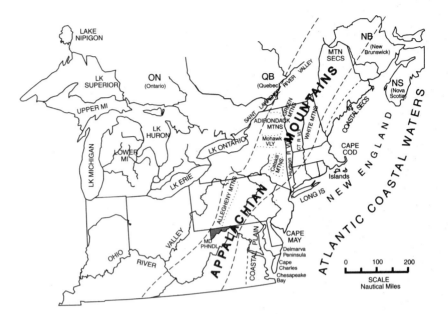

Fig. D-4. Common geographical area designators (northeast).

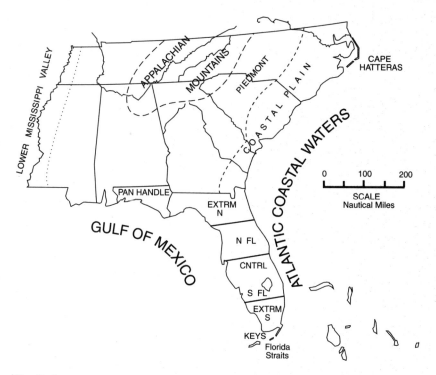

Fig. D-5. Common geographical area designators (southeast).

above ground level (AGL)   A cloud layer measured from the surface to the cloud base.

absolute instability   A condition of the atmosphere where vertical displacement is spontaneous, whether saturated or unsaturated.

absolute stability   A condition of the atmosphere that resists vertical displacement whether a parcel is saturated or unsaturated.

accretion   The deposit of ice on aircraft surfaces in flight as a result of the tendency of cloud droplets to remain in a liquid state at temperatures below freezing.

adiabatic process   A thermodynamic change of state with no transfer of heat or mass across the boundaries, where compression always results in heating and expansion results in cooling.

advection   The horizontal transport of an atmospheric property.

aeronautical decision making   The ability to obtain all available, relevant information, and evaluate alternative courses of action, then analyze and evaluate their risks and determine the results.

air mass   A widespread body of air whose homogeneous properties were established while that air was over a particular region of the earth's surface, and that undergoes specific modifications while moving away from its source region.

AIRMET Bulletin   An inflight weather advisory program intended to provide advance notice of potentially hazardous weather to small aircraft.

altimeter setting   A value of atmospheric pressure used to correct a pressure altimeter for nonstandard pressure.

anomalous propagation (AP)   Under highly stable atmospheric conditions (typically on calm, clear nights), the radar beam can be refracted almost directly into the ground at some distance from the radar, resulting in an area of intense, but false, echoes.

anticyclonic   Having a clockwise rotation in the northern hemisphere, associated with the circulation around an anticyclone (high-pressure area).

arc cloud   See *arc line.*

arc line   An arc-shaped line of convective clouds often observed in satellite imagery moving away from a dissipating thunderstorm area.

arrow echo   Severe storms are tilted though the atmosphere, which allows them to be steady state and often severe. This tilting is sometimes indicated on radar by an arrow-shaped echo.

asymmetric echo   Like the arrow echo, indicates a tilted storm, with its associated severe weather. The storm produces echo shapes and colors that are not even or concentric.

atmospheric phenomena   As reported on METAR, atmospheric phenomena are weather occurring at the station and any obstructions to vision. Obstructions to vision are reported only when the prevailing visibility is less than 7 miles.

atmospheric property   A characteristic trait or peculiarity of the atmosphere such as temperature, pressure, moisture, density, and stability.

augmented   In reference to a surface weather observation, the term means someone is physically at the site monitoring the equipment and has overall responsibility for the observation.

AUTO   When used in METAR and SPECI, or a radar report, indicates the report comes from an automated observation.

automated observation   An automated report indicates the observation was derived without human intervention.

attenuation   Any process that reduces the power density within the radar beam. A target or obstruction close to the antenna may absorb and scatter so much of the energy that little passes to a more distant target.

Automatic Terminal Information Service (ATIS)   A recorded service provided at tower-controlled airports to provide the pilot with weather, traffic, and takeoff and landing information.

automated observation   An automated report indicates the observation was derived without human intervention.

Automated Surface Observing System (ASOS)   A computerized system similar to AWOS, but developed jointly by the FAA, NWS, and Department of Defense; in addition to standard weather elements the system encodes climatological data at the end of the report.

Automated Weather Observing System (AWOS)   A computerized system that measures sky condition, visibility, precipitation, temperature, dew point, wind, and altimeter setting. It has a voice synthesizer to report minute-by-minute observations over radio frequencies, telephone lines, or local displays.

baroclinic   A state of the atmosphere where isotherms—lines of equal temperature—cross contours, temperature and pressure gradients are steep, and temperature advection takes place. A baroclinic atmosphere enhances the formation and strengthens the intensity of storms. It is characterized by an upper-level wave one-quarter wavelength behind the surface front.

barotropic   Barotropic is an absence of isotherms, or the opposite of baroclinic. Theoretically, an entirely barotropic atmosphere would yield constant-pressure charts with no height or temperature gradients or vertical motion.

base reflectivity   A radar image currently available from the NWS Internet site with the radar's antenna tilted 0.5° above the horizon. It is a sample of only one elevation within the radar's range.

beam resolution   The ability of a radar to distinguish between targets at the same range, but at different azimuths.

blocking high   An upper-level area of high pressure that blocks approaching weather systems. See *omega block*.

boundaries Zones in the lower atmosphere characterized by sharp gradients or discontinuities of temperature, pressure, or moisture and often accompanied by convergence in the wind field. Examples include surface fronts, dry lines, and outflow boundaries. In the last case, the boundary is produced by a surge of rain-cooled air flowing outward near the surface from the originating area of convection. In an unstable air mass, thunderstorms tend to develop along these zones and especially at intersections of two or more boundaries.

bow echo A radar echo typically associated with fast-moving, broken, or solid lines of thunderstorms. Severe weather will most likely develop along the bulge and at the northern end of the echo pattern.

BWER/WER/LEWP Bounded Weak Echo Region/Weak Echo Region/Line Echo Wave Pattern. All of these weather radar terms are indicators of strong thunderstorms and the development of severe weather.

ceiling (CIG) A broken or overcast layer measured from the surface to the cloud base.

Celsius A temperature scale where 0° is the melting point of ice and 100° the boiling point of water.

circular polarization (CP) An ATC radar circuit that reduces the radar's sensitivity to light and moderate precipitation.

clear air mode This NWS radar mode has the slowest antenna rotation rate that permits the radar to sample a given volume of the atmosphere longer. Increased sampling increases the radar's sensitivity and ability to detect smaller objects than is possible in precipitation mode.

clear air turbulence (CAT) Nonconvective wind shear turbulence occurring at or above 15,000 ft, although it usually refers to turbulence above 25,000 ft.

closed low An area of low pressure aloft completely surrounded by a contour.

closed cell stratocumulus Common over water, this is the satellite-viewed oceanic stratocumulus associated with an inversion. It is associated with high pressure.

cloud band A nearly continuous cloud formation with a distinct long axis, a length-to-width ratio of at least 4 to 1, and a width greater than 1° of latitude (60 nm).

cloud element   The smallest cloud form that can be resolved on satellite imagery from a given satellite system.

cloud line   A narrow cloud band in which individual elements are connected and the line is less than 1° of latitude in width. Indicates strong winds, often 30 knots or greater over water.

cloud shield   A broad cloud formation that is not more than 4 times longer than it is wide. Often it is formed by cirrus clouds associated with a ridge or the jet stream.

cloud streets   A series of aligned cloud elements that are not connected. Several cloud streets usually line up parallel to each other, and each street is not more than 10 miles wide.

cold-air advection   In the lower atmosphere a downward, stabilizing vertical motion producer; in the upper atmosphere an upward, destabilizing vertical motion producer.

cold-core low   A low-pressure area that intensifies aloft. When this type of low contains closed contours at the 200-mb level, its movements tend to be slow and erratic.

cold pool   Generally refers to an area at 500 mb in which the air temperature is colder than adjacent areas. Other atmospheric conditions being equal, thermodynamic instability is greater beneath cold pools, thus making thunderstorm development more likely.

comma cloud system   A cloud system that resembles the comma punctuation mark. The shape results from differential rotation of the cloud border and upward and downward moving air.

comma head   The rounded portion of the comma cloud system. This region often produces most of the steady precipitation.

comma tail   The portion of the comma cloud that lies to the right of, and often nearly parallel to, the axis of maximum winds.

composite radar display   One picture or display obtained from two or more individual radar sites.

composite reflectivity   This radar image displays maximum echo intensity from any elevation angle at every range from the radar. This product is used to reveal the highest reflectivity in all echoes.

cone of silence   When used with radar, is the area directly over the antenna where radar echoes are not detected.

conditional instability   A condition of the atmosphere where a parcel will spontaneously rise as a result of becoming saturated when forced upward.

conduction   The process of transferring energy by means of physical contact.

confluence   A region where streamlines converge. The speed of the horizontal flow will often increase where there is confluence. It is the upper-level equivalent of surface convergence.

contours   Lines that connect areas of equal height on constant-pressure charts.

Constant-pressure surface   A surface where atmospheric pressure is equal—the height of the surface changes with pressure changes, but the pressure itself remains constant.

convection   The vertical transport of an atmospheric property.

convergence   Air flowing together near the surface is forced upward because of convergence. It is a vertical motion producer that tends to destabilize the atmosphere near the surface. An inward flow or squeezing of the air.

convergence zone   An area where surface convergence is occurring, often associated with a large body of water.

Coordinated Universal Time   Formerly Greenwich Mean Time, also known as Z or ZULU time, Coordinated Universal Time (UTC) is the international time standard. UTC is used in all aviation time references.

cumuliform   Cumuliform describes clouds that are characterized by vertical development in the form of rising mounds, domes, or towers, and an unstable air mass.

cutoff low   See *closed low.*

cyclogenesis   The development or strengthening of a cyclone.

cyclonic   Having a counterclockwise rotation in the northern hemisphere, associated with the circulation around a cyclone or low-pressure area.

decibels (dBZ)   NEXRAD radars display radar intensity returns as energy with units in dBZ.

deformation zone   An area within the atmospheric circulation where air parcels contract in one direction and elongate in the perpendicular direction. The narrow zone along the axis of elongation is called the deformation zone. Deformation is a primary factor in frontogenesis and frontolysis.

dendritic pattern   This branchy sawtooth pattern identifies areas of snow cover. Mountain ridges above the tree line are essentially barren and snow is visible; in the tree-filled valleys, most of the snow is hidden beneath the trees.

density   The weight of air per unit volume.

density altitude   Density altitude is pressure altitude corrected for nonstandard temperature.

dew point   The temperature to which air must be cooled, water vapor remaining constant, to become saturated.

dew point front   See *dry line*.

diabatic   A process that involves the exchange of heat with an external source; also called nonadiabatic. The loss may occur through radiation resulting in fog or low clouds, or conduction through contact with a cold surface.

difulence   The spreading apart of adjacent streamlines. The speed of horizontal flow often decreases with a difluent zone. It is the upper air equivalent of surface divergence and activates or perpetuates thunderstorm development.

dig or digging   Indicates a trough with a strong southerly component of motion. These troughs contain considerable strength and are difficult to forecast accurately.

Direct User Access Terminal (DUAT)   A computer terminal where pilots can directly access meteorological and aeronautical information, and file a flight plan without the assistance of an FSS.

divergence   Subsiding air diverges, or spreads, at the surface. Divergence is a downward motion producer that tends to stabilize the atmosphere near the surface.

dry line   An area within an air mass that has little temperature gradient, but significant differences in moisture. The boundary between the dry and moist air produces a lifting mechanism. Although not a true front, it has the potential to produce hazardous weather. It is also known as a *dew point front*.

dry slot   A satellite meteorology term used to describe a cloud feature associated with an upper-level short-wave trough. Generally speaking, the cloud system is shaped like a large comma. As the system develops, sinking air beneath the jet stream causes an intrusion of dry, relatively cloud-free air on the upwind side of the comma cloud. The air of the intrusion is the dry slot. It is commonly the location where lines of thunderstorms subsequently develop.

embedded thunderstorm   A thunderstorm that occurs within nonconvective precipitation. A thunderstorm that is hidden in stratiform clouds.

enhanced cumulus   Towering cumulus as seen on satellite imagery; cumulus with vertical development without cirroform tops indicated by texturing and shadowing.

enhanced infrared (IR) image   A process by which infrared imagery is enhanced to provide increased contrast between features to simplify interpretation. This is done by assigning specific shades of gray to specific temperature ranges.

enhanced V   A cloud top signature sometimes seen in enhanced infrared imagery in which the coldest cloud top temperatures form a V shape. Storms that show this cloud top feature are often associated with severe weather.

eye wall   The area of thunderstorms that surrounds the eye of a tropical storm.

Fahrenheit   A temperature scale where 32° is the melting point of ice and 212° the boiling point of water.

fine line/thin line   At times weather radar picks up dust or debris that appear as a fine or thin line caused by a dry front or gust front. It indicates the presence of low-level wind shear.

finger echo   Like a hook echo, it represents a strong probability of severe thunderstorms and tornadoes.

flight level   Pressure altitude read off an altimeter set to standard pressure of 29.92; altitude used in the United States above 17,999 ft.

freezing level   As used in aviation forecasts, the level at which ice melts.

front   A boundary between air masses of different temperatures, moisture, and wind.

frontal zone   See *front*.

frontogensis   The process by which frontal systems are formed.

frontolysis   The process of frontal system dissipation.

Geostationary Operational Environmental Satellite   See *GOES*.

GOES   Geostationary Operational Environmental Satellites, normally located about 22,000 nm above the equator at 75° W and 135° W. The satellites provide half-hourly visible and infrared imagery.

Great Basin   The area between the Rockies and Sierra Nevada Mountains, consisting of southeastern Oregon, southern Idaho, western Utah, and Nevada.

ground clutter   Interference of the radar beam due to objects on the ground.

gust front   A low-level windshift line created by the downdrafts associated with thunderstorms and other convective clouds. Acting like a front, these features might produce strong gustiness, pressure rises, and low-level wind shear.

hail shaft   A shaft of hail detected on weather radar.

hook echo   A bona fide hook echo indicates the existence of a mesolow associated with a large thunderstorm cell. Such mesolows are often associated with severe thunderstorms and tornadoes. Hook echoes are not seen on ATC radars.

impulse   A weak, mid- to upper-level and fast-moving short wave feature that can kick off thunderstorms.

inches of mercury   For aviation purposes we commonly relate atmospheric pressure to inches of mercury (in Hg) for altimeter setting.

infrared   Satellite imagery that measures the relative temperature of clouds or Earth's surface.

instrument flight rules (IFR)   A set of rules governing the procedures for conducting instrument flight. Also a term used by pilots and controllers to indicate type of flight plan.

instrument meteorological conditions (IMC)   Meteorological conditions expressed in terms of visibility, distance from cloud, and ceiling less than the minimums specified for visual meteorological conditions.

intensity level   Used to describe radar precipitation intensity on RAREPs (SD) and the Radar Summary Chart.

intermountain region   The area of the western United States, west of the Rocky Mountains and east of the Sierra Nevada Mountains, that includes Idaho and Arizona.

international standard atmosphere (ISA)   A hypothetical vertical distribution of atmospheric properties (temperature, pressure, and density). At the surface, the ISA has a temperature of 15°C (59°F), pressure of 1013.2 mb (29.92 in), and a lapse rate of approximately 2°C in the troposphere.

intertropical convergence zone (ITCZ)   The dividing line between the southeast trade winds and the northeast trade winds of the southern and northern hemispheres, respectively.

inversion   A lapse rate where temperature increases with altitude.

isobars   Lines connecting equal values of surface presssure.

jet streaks   See *jet stream.*

jet stream   A segmented band of strong winds that occur in breaks in the tropopause.

jetlets   See *jet stream.*

lapse rate   The decrease of an atmospheric variable with height, usually temperature.

level of free convection (LFC)   The point where a parcel becomes saturated and upward movement becomes spontaneous.

LEWP   See *BWER/WER/LEWP.*

lifted condensation level (LCL)   The level, or altitude, where a lifted parcel becomes saturated.

lightning   A meteorological condition in thunderstorms caused by differences in electrical charge that produce a visible electrical discharge.

liquid water content (LWC)   The total mass of water contained in all the liquid cloud droplets within a unit volume of cloud. Units of LWC are usually grams of water per cubic meter of air (g/nm^3).

location identifier   Consisting of three to five alphanumeric characters, location identifiers are contractions used to identify geographical locations, navigational aids, and intersections.

LOCID   See *location identifier.*

Loop   A series of satellite or radar images, displayed in sequence, to obtain a trend of weather conditions.

low   An area of low pressure completely surrounded by higher pressure.

low-level wind shear (LLWS)   Wind shear that occurs within 2000 ft of the surface.

macroburst   A downburst that affects a path longer than 2 nm and may persist for up to 30 minutes.

mean sea level (MSL)   A cloud layer or altitude measured from average sea level to the cloud base, or altitude or elevation.

mean storm motion vector   The mean wind vector is the direction and magnitude of the mean winds from 5000 ft AGL to the tropopause. It can be used to estimate cell movement.

mean wind vector   See *mean storm motion vector.*

melting level   The temperature at which ice melts, often referred to as the *freezing level.*

mesocyclone   A stage in the development of a tornado ranging from 1.5 to 4.5 nm in diameter.

mesolow   Also known as *mesocyclone,* a mesolow is a small area of low pressure within a severe thunderstorm. Tornadoes can develop within the vortex.

mesoscale   Small-scale meteorological phenomena that can range in size from that of a single thunderstorm to an area the size of the state of Oklahoma.

mesoscale convective complex (MCC)   A large, organized, homogeneous convective weather system that can cover an area the size of several states. MCCs tend to form during the morning hours.

METAR   Meteorological Aviation Routine surface weather report.

microburst   A small-scale, severe storm downburst less than $2^1/_2$ miles across. Reaching the ground, the burst continues as an expanding outflow producing severe wind shear.

microscale   See *subsynoptic scale.*

millibar   For aviation purposes, we commonly relate atmospheric pressure to inches of mercury or millibars (mb).

moving target indicator (MTI)   An ATC radar circuit that displays only moving targets.

NAVAID   See *navigational aid.*

navigational aid   An electronic, ground- or space-based, device used in aerial navigation.

negative vorticity advection   Area of low values of vorticity producing downward vertical motion.

neutral stability   An atmospheric condition in which a parcel, after being displaced, remains at rest, even when the displacing force ceases.

NEXRAD   The next-generation doppler weather radar system.

omega block   A blocking high that on weather charts resembles the Greek letter omega.

open cell   On satellite imagery, a pattern of clear air surrounding individual convective cells.

Operation and Supportable Implementation System (OASIS) Equipment designed to replace existing FSS Model 1 computer and graphics system.

orographic   A term used to describe the effects caused by terrain, especially mountains, on the weather.

outflow boundary   An outflow boundary is a surface boundary left by the horizontal spreading of thunderstorm-cooled air. The boundary is often the lifting mechanism needed to generate new thunderstorms.

overrunning   A condition in which airflow from one air mass is moving over another air mass of greater density. The term usually applies to warmer air flowing over cooler air as in a warm frontal situation. It implies a lifting mechanism that can trigger convection in unstable air.

overshooting tops   On satellite imagery, it refers to the tops of severe thunderstorms, which often develop above the normal cloud tops as a result of the severe updrafts of the storm.

p-static   See *precipitation static.*

parcel   A small volume of air arbitrarily selected for study; it retains its composition and does not mix with the surrounding air.

pendant echo   A pendant represents one of the most severe storms— a supercell.

polar front   A semipermanent, semicontinuous front separating air masses of tropical and polar origin.

polar jet   The jet stream located at the break between the polar tropopause and subtropical tropopause.

polar orbiters   Meteorological satellites that orbit along a north-south axis crossing Earth's poles at an altitude from about 400 to 600 nm.

positive vorticity advection (PVA)   Areas of high positive values of vorticity producing upward vertical motion.

precipitation   Any or all of the forms of water particles, whether liquid or solid, that fall from the atmosphere and reach the ground.

precipitation mode   This NWS radar mode is designed specifically to detect precipitation size returns.

precipitation static   Low-frequency radio communication and navigation interference caused by corona discharge from radio antennas or other protuberances on an aircraft.

pressure   Pressure is force per unit area.

pressure altitude   The altitude above the mean-sea-level constant-pressure surface; indicated altitude with the altimeter set to 29.92 inHg.

probability   The ratio of the chances favoring an event to the total number of chances for and against it.

radiation   The process of transferring energy through space without the aid of a material medium.

range attenuation   The loss of radar power density due to distance from the antenna.

RAREP   Automated, textual radar weather report.

relative humidity   The ratio, expressed as a percentage, of water vapor present in the air compared to the maximum amount the air could hold at its present temperature.

ridge   An elongated area of high pressure.

St. Elmo's fire   An electrical phenomenon associated with thunderstorms causing a corona discharge.

sensitive time control (STC)   A radar circuit used to compensate for range attenuation of echoes, which is loss of power density due to distance from echoes. STC-displayed intensity remains independent of range; therefore, targets with the same intensity, at different ranges, appear the same on the display.

scalloped echo   Scallop-shaped echoes indicate turbulent motion within the cloud. There is a good probability of hail associated with these echoes.

SD (storm detection)   See *RAREP.*

severe thunderstorm   A thunderstorm that produces winds of 50 knots or more, or hail $3/4$ inch in diameter or greater.

severe wind shear   See *wind shear.*

shear axis   An axis indicating maximum lateral change in wind direction, as in an elongated circulation. This lateral change or shear might be either cyclonic or anticyclonic.

short wave With wave lengths shorter than long waves, they tend to move rapidly through the long wave circulation. They can intensify or dampen weather systems.

showers Showers are characterized by the suddenness with which they start and stop, and rapid changes of intensity.

SIGMET A significant meteorological advisory that warns of phenomena that affect all aircraft.

slant range visibility The visibility between an aircraft in the air and objects on the ground.

SPECI A special surface aviation weather report.

squall line An organized line of thunderstorms.

snow A porous, permeable aggregate of ice grains that can be predominately single crystals or close groupings of several crystals.

stability The property of an air mass to remain in equilibrium; its ability to resist displacement from its initial position.

standard atmosphere See *international standard atmosphere.*

storm detection equipment Refers to airborne access to real-time weather radar or lightning detection equipment.

stratiform Stratiform describes clouds of extensive horizontal development and a stable air mass.

stratosphere The layer of the atmosphere above the tropopause.

subsynoptic scale Small, microscale events that often are not within the detection system currently available—tornadoes and microbursts.

sublimation The process by which ice changes directly to water vapor, or water vapor directly to ice. The sublimation of ice to vapor is a cooling process, and water vapor to ice is a warming process.

subsidence Downward vertical motion of the air.

supercell A thunderstorm where updrafts and downdrafts coexist, prolonging the life of the cell and often producing a severe thunderstorm and tornadoes.

supercooled Refers to liquid water or water vapor that exists at temperatures below freezing.

synoptic scale Large-scale weather patterns the size of the migratory high- and low-pressure systems of the lower troposphere with wave lengths on the order of 1000 miles.

TAF Terminal Aerodrome Forecast.

temperature   A measurement of the average speed of molecules.

terminator   The sunset line seen on visible satellite imagery.

texture   On satellite imagery, texture refers to the lumpy, rounded, billowy, and puffy appearance of clouds.

towering cumulus   Growing cumulus that resembles a cauliflower, but with tops that have not yet reached the cirrus level.

thin line   See *fine line/thin line.*

transverse cirrus banding   Irregularly spaced bandlike cirrus clouds that form nearly perpendicular to a jet stream axis. They indicate turbulence associated with the jet.

tropical cyclone   A general term applied to any low-pressure area that originates over tropical oceans.

tropopause   The boundary between the troposphere and the stratosphere, typically an isothermal layer.

troposphere   The lower layer of Earth's atmosphere containing about three-quarters of the atmosphere by weight.

trough   An elongated area of low pressure.

tule fog   *Tule* [tôô' lê] is a Spanish word for bulrushes, a marsh plant that grows during the winter season in California's Central Valley; it describes an extensive area of radiation fog that develop in this area.

upslope   The orographic effect of air moving up a slope, which tends to cool adiabaticly.

V notch   A large, isolated echo that sometimes has the configuration of a V or U shape. A V notch often accompanies severe thunderstorms and tornadoes.

vertical motion   Upward or downward motion in the atmosphere.

Video Integrator and Processor (VIP)   Prior to the commissioning of the NEXRAD network, weather radar intensity levels were described as VIP-levels by using a six-level scale to indicate precipitation intensity.

visible moisture   Moisture in the form of clouds or precipitation.

visual flight rules (VFR)   Rules that govern flight in visual meteorological conditions.

visual meteorological conditions (VMC)   Meteorological conditions expressed in terms of visibility, distance from cloud, and ceiling equal to or better than specified minima.

vort lobe A contraction for vorticity lobe. It usually applies to the 500-mb level and identifies an area of relatively higher values of vorticity. It is synonymous with short-wave trough or upper-level impulse. Generally speaking, there is rising air ahead of the vort lobe and sinking air behind.

vort max A contraction for vorticity maximum. It usually applies to the 500-mb level and refers to a point along a vortex. In the most general use, any flow possessing vorticity; indicates a circulation or rotation within the atmosphere.

wall cloud See *eye wall*.

warm-air advection A condition in the atmosphere characterized by air flowing from a relatively warmer area to a cooler area. It is often accompanied by upward vertical motion that in the presence of sufficient instability leads to thunderstorm development. In the upper atmosphere, a downward, stabilizing vertical motion producer; in the lower atmosphere, an upward, destabilizing vertical motion producer.

water vapor The invisible water molecules suspended in the air.

wave A pattern of ridges and troughs in the horizontal flow as depicted on upper-level charts. At the surface, a wave is characterized by a break along a frontal boundary. A center of low pressure is frequently located at the apex of the wave.

Weather Fixed Map Unit (WFMU) ARTCC radar circuit that allows the controller to display two intensities of precipitation echoes.

Weather and Radar Processor (WARP) WARP generates a NEXRAD mosaic overlay on ATC controller's displays.

WER See *BWER/WER/LEWP*.

whiteout An atmospheric optical phenomenon in which the pilot appears to be engulfed in a uniformly white glow. Neither shadows, horizon, nor clouds are discernible; sense of depth and orientation is lost.

wind shear Any rapid change in wind direction or velocity. Low-level wind shear (LLWS) is generally shear that occurs within about 2000 ft of the surface. LLWS is classified severe when a rapid change in wind direction or velocity causes an airspeed change greater than 15 knots or vertical speed change greater than 500 ft/min.

zonal flow A wind flow that is generally in a west to east direction.

Index

About the Author

Terry T. Lankford is a retired FAA Weather Specialist, now being utilized as a consultant by NASA and the National Weather Service on flight weather training and techniques. He holds multiple current flight certifications, including that of flight instructor. Mr. Lankford is also the author of five previous books for McGraw-Hill.